The Feminine Monarchie

or

THE HISTORY OF BEES

By

Charles Butler
1623

Edited with an introduction and notes by John Owen

Northern Bee Books

The Feminine Monarchie or THE HISTORY OF BEES

Edited by John Owen
© John Owen, 2017

ISBN 978-1-904846-04-8

Published by Northern Bee Books, 2017
Scout Bottom Farm
Mytholmroyd
Hebden Bridge HX7 5JS (UK)

Design and artwork: D&P Design and Print

Printed by Lightning Source, UK

The Feminine Monarchie

or

THE HISTORY OF BEES

showing their admirable nature and properties,
their generation and colonies, their government, loyalty, art,
industry, enemies, wars, magnanimity, together with the
right ordering of them from time to time:
And the sweet profit arising thereof.
Written out of Experience
By

Charles Butler
1623

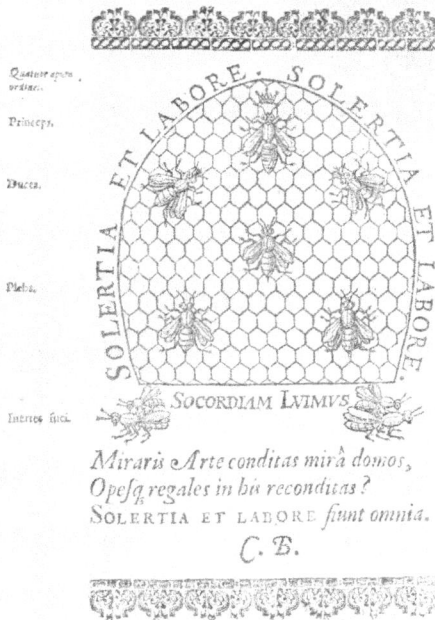

Quàtuto apte
ordine.

Princeps.

Ducca.

Plebs.

Inertes fuci.

SOLERTIA ET LABORE · SOLERTIA ET LABORE

SOCORDIAM LVIMVS

Miraris Arte conditas mirâ domos,
Opesq regales in bis reconditas?
SOLERTIA ET LABORE *fiunt omnia.*

C. B.

Introduction:
Charles Butler (1571 - 1647)

Do you want to keep bees? asks Charles Butler in 1609. Well, place the hives in your garden where you can keep an eye on them and your own flowers will support their foraging. When the hives start increasing, as they will, put them out into fenced-off areas, on the south side of the house, where they're protected. In the winter, when the snow lies thick on the ground – this was England before climate change – bring them into sheds where there are no cattle. Make a little cover from wheat or rye straw, *a hackle*, and the chances are that you will be able to keep them safe until the reviving warmth of spring arrives.

Butler called bees 'delightful, profitable and necessary creatures' although he didn't keep his out of intellectual curiosity alone. He kept them for profit, having experimented earlier, and possibly not very successfully, with raising silk worms[1]. A large dowry of £400 was raised for the marriage of one of his daughters, Elizabeth, in 1633[2], through the sale of honey and wax, and it is quite likely that Butler was able to partly finance the printing of his several books by this cottage industry, based at his Hampshire vicarage.

Skeps similar to those in use in the seventeenth century;
the Weald and Downland Open Air Museum, West Sussex;

The Feminine Monarchie is not only a bee keeping manual. It is also one of the earliest examples in English of natural history writing. Its author is fascinated by bees: their nature, their reproduction, the way they organise their colony and especially, their government by a female monarch. In precise and intelligent prose, based on years of close observation of the activity in domed skep hives, Charles Butler describes how best to manage bees so that not only will they produce 'sweet profit' in the form of honey and wax, but also how they will prove to be a 'delightful recreation' with just a little care and attendance. In passing, the sights, smells and sounds of seventeenth century rural England are vividly brought to life, together with some of its plants, animals and insects. Butler's quiet but authoritative voice initiates the reader into the world of bees which his observations have discovered. While he pays lip service to the views of classical authors, he is thoroughly in tune with the emerging scientific spirit of the seventeenth century in which deductions about animal and insect behaviour are made on the basis of evidence and observation, rather than religious dogma or romantic allusions.

It's a remarkable achievement, and is one of the reasons why the book has never entirely fallen out of fashion, especially among beekeepers who have long appreciated the author's practical and theoretical knowledge, and who have given to Charles Butler the title 'father of English bee-keeping'.

Butler was resigned to thinking that his book, *The Feminine Monarchie,* would not be widely read[3] on its publication in 1609, when he was most probably thirty eight years[4] of age. The sixteenth century saw an explosion of printed books[5] and Butler's *Feminine Monarchie* benefitted from the arrival of this new technology. A scholarly clergyman and schoolteacher, he published extensively, and had wide ranging interests in grammar, logic and music. It remains a conundrum why, with his many gifts, Butler remained a country parson for all his adult life.

Wootton St Lawrence parish, in the county of Hampshire, where he was vicar from 1600 until his death in 1647, was a thinly populated and poorly endowed parish. Over seventy years after Butler's death, its population is recorded as being '290 souls'[6], including a small number of gentry. It was in this inauspicious context that Butler (like the later Gilbert White of Selbourne, whose rural isolation was similar) managed to sustain a scholarly and productive life, while carrying out his priestly duties. Later writers noted with sympathy the lack of material reward that seemed to attend him. Wootton St Lawrence, thought Anthony Wood[7], was but 'a poor preferment, God wot, for so worthy a scholar'. Samuel Johnson[8] described him as 'a man who did not want an understanding which might have qualified him for better employment'. Thomas Fuller admired him for his piety and felt that his several books were 'the receptacle of the leakage and superfluities of his study' [9].

His Life

Butler was born in High Wycombe, most probably in 1571 and is likely to have benefitted (like William Laud) from a grammar school education, in which Latin and Greek formed a key part.

Butler may have attended the Royal Grammar School in High Wycombe, housed then in St John's Hospital, the ruins of which are in the town. Photo Paul Hurley

His intelligence and ability was noted, and he became a boy chorister at Magdalen College, Oxford, at the age of eight. Effectively this was a way for poorer boys to prepare for university entrance; at the age of 10, Butler matriculated, taking his BA in 1584 and his MA in 1587. After leaving the university he became rector of a small parish at Nately Scures, near Basingstoke, Hampshire, in 1593 and two years after, became master of the Holy Ghost School in Basingstoke.

Nately Scures Church

In 1600 Butler resigned from both offices in order to move to Wootton St Lawrence (three miles from Basingstoke), where he remained as incumbent for the rest of his life. He lived through the reigns of three monarchs, and spent the most formative years of his life in Tudor England, under Elizabeth 1. *The Feminine Monarchie* first appeared six years after Elizabeth's death, in which the political significance of the book's title would not have been overlooked. He died in a very different England at a time of great political and social turmoil, during the years of the English Civil War. Unlike many of his contemporaries, Butler was not ejected from his living and was allowed to remain in his parish until his death, probably out of respect for his age and learning. His neatly written entries for baptisms, weddings and funerals in the well maintained registers of Wootton St Lawrence can still be viewed, together with entries in the Churchwardens' Accounts.

Wootton St Lawrence Church, where Butler was vicar for 47 years

Charles Butler was married; his wife died in 1628, though her name is not known. He had three sons, William, Edmund and Richard, and two daughters, Elizabeth and Briget; the latter died in infancy. *The Feminine Monarchie* makes reference to the practical assistance he received in managing his bees, and it is likely on these occasions he is referring to the involvement of his wife and children.

Butler died on 29 March 1647 and was buried in an unmarked grave in the chancel of his church at Wootton St Lawrence. That there was no inscription to mark the place may have been in keeping with his wishes, as was the case with George Herbert, buried at Bemerton in 1633. 'He lies in the chancel, under no large, nor yet very good, marble gravestone, without any inscription...' [10] but in Butler's case it is more likely a reflection of his relative poverty. In 1652 a parliamentary minister, a plain Mr Dennys, was appointed to the charge of the parish, in line with the Rump Parliament's attempts to control the appointment of clergy, and allow approved ministers only to preach in the parishes.

Butler's signature on the 14th century parchment inner cover of the Churchwardens' Accounts at Wootton

His published writings

Charles Butler had a wide range of scholarly interests and pursuits, which included logic, music, English grammar, apiculture and ecclesiastical canon law. He was therefore much more than being solely a beekeeper. H M Fraser, who assembled copious notes and research on Butler's life and came closest to anyone in the last century in rehabilitating Butler's intellectual and practical reputation, observed that 'a biographical study of this versatile author is long overdue, probably because it would be difficult for one biographer to do justice to all his books' [11]. *The Feminine Monarchie* has stood the test of time, and has, in the last 400 years been re-printed and referred to more than Butler's other books. His works include:

The Logic of Ramus (1597)

The Feminine Monarchie, or History of Bees (1609, 1623; 1634 in Butler's reformed spelling; 1673 a Latin translation; 1704)

The Marriage of Cousins (1625)

The English Grammar (1633)

The Principles of Music (1636)

Significance of 'The Feminine Monarchie'

The Feminine Monarchie is an early and substantial contribution to the literature of apiculture. There were other beekeeping books of that period, such as Edmund Southerne's *'A treatise concerning the right use and ordering of bees'* (1593) and John Levett's *'The ordering of bees'*, published in 1634 but most probably written before Butler's book[12]. These books are of a different quality to Butler's writing in that they lack the organisation and scholarship displayed by Butler. *The Feminine Monarchie* has ten chapters, coherently arranged, complete with numbered margin notes which give the sources of quotations; reasons are supplied for assertions and statements in the main text. The title page proclaims 'written out of experience by Charles Butler' and this reliance on observed experience is repeated throughout the text. Butler occasionally falls back on legendary material when he is uncertain of their provenance – for example, the Russian bear who unwittingly hauls a man out of a lake of wild forest honey (chapter 6) and the thief who is justly punished by the bees for stealing St Medard's honey (chapter 7). Yet these are not characteristic of Butler's approach, which in the main is evidence based, analytic, and rational. He is a naturalist before he is a classicist. He quotes the Roman authors, including Aristotle (widely regarded as the classical authority on bees at the time) but is quick to dismiss their views if he finds these are not verified by his own experience. The authority of received wisdom is pitted against the new authority of observational evidence – and the authority of experience wins.

All this Butler achieves without the benefit of microscopical study, which was not widely available until later in the seventeenth century[13] but even then would not have been easily accessed by a man of Butler's limited means. Jan Swammerdam (1637 – 1680), Dutch biologist and microscopist was able to take advantage of the new invention and it was he, not Butler, who was able to show conclusively that the 'king' bee, long recognised by the classical writers, has ovaries, and is the actual 'mother' of the entire colony.

While Butler's book and his methods do not create any new skep equipment, or revolutionary understandings about bees, he does provide a wealth of insight into the ways in which the Amazonian queen distributes her power throughout the hive. It is in this context that Butler's book acquires a political dimension, to which recent writers have been sensitive[14]. Butler was a follower of Archbishop Laud, with whom he was contemporary at Oxford. (Laud's career followed a different trajectory to Butler's. Finding himself after university in a rural parsonage in Cuckstone, Kent in 1610[15] Laud decided he could not face the prospect of such rural isolation for the rest of his life, and promptly engineered his return to metropolitan, university life.) Both men believed in the vital importance of monarchy for stable and enduring government. As archbishop, Laud encouraged a uniformity of liturgy in the English church and this brought him into conflict with Puritan opinion; he encouraged scholarship amongst his clergy, and in the parish church made the communion table rather than the pulpit its centre. 'Laudian frontals' were so named following Laud's encouragement of such drapes on the communion table. At Wootton St Lawrence, the Churchwardens' Accounts during Butler's time as vicar indicate that he oversaw similar changes to the furniture of the chancel[16]. So when we read in *The Feminine Monarchie* passages like this:

...the bees abhor government by many, as well as anarchy, God having showed in them to men, an express pattern of a perfect monarchy, the most natural and absolute form of government...' (chapter 1)

it is difficult to avoid the conclusion that Butler is making a plea to his fellow citizens in the agitated years which preceded the beheading of the monarch and the English Revolution of the 1640s.

Butler skilfully declares his allegiance to the government of the bees, but not, at least explicitly, to the Stuart monarchy in place at the time the second edition of his book appears. It was most probably this ability to keep a low profile which allowed him to avoid ejection from his living in 1642, at the outbreak of civil war.

To his contemporary readers, at least the more fleeting ones, his book is about bees, not politics, and in this arena he is certainly knowledgeable, employing a specific vocabulary relating to skep management, much of which has now become obsolete, and makes the reading of *The Feminine Monarchie*, in its original form, a ponderous task. Hence the need for a new edition.

This edition

This edition has been edited with the aim of realising Butler's original intention of producing a helpful and readable text on beekeeping. Obsolete words and expressions have been replaced by current phrases. Butler, in the writing style of his day, tends to use several clauses in long sentences, the longer of which I have broken into shorter sentences for the sake of clarity.

I have omitted the long excerpts in Latin from the classical authors whom Butler cites. For anyone who wants to refer to them, they are easily found by reference to the full edition. I've also omitted Butler's extensive margin notes and references, because they primarily reinforce the argument in his text, and again, can be read in the original edition. I have added a glossary of obsolete words, together with footnotes where I have felt some further commentary would be helpful. Also included is an index of key words and topics which adds to the usefulness of the critical apparatus in this edition.

I have also omitted the Preface to the Reader, and the original dedicatory verses in Latin by George Wither, of Manydown Park, in Wootton.

The use of the astronomical months which Butler thought are 'most natural and fitting to my purpose' for 'distinguishing the times of year' is retained. A table at the back of the book is given for reference.

The text of the 1623 edition of *The Feminine Monarchie* is used, rather than the 1609 edition, because the famous Bees' Madrigal appears in the later version.

It seems appropriate that this edition of *The Feminine Monarchie* is being prepared in the year in which Queen Elizabeth II has reached ninety years of age.

Part of the memorial window to Charles Butler, in Wootton St Lawrence Church, to mark the Coronation of Queen Elizabeth II, 1953

1   Armstrong, Patrick *The English Parson-Naturalist*, 2000, Gracewing, p 95

2   Aubrey, John, *Brief Lives*, quoted by A S Bates, *Charles Butler, Vicar of Wootton*, church pamphlet

3   'I am out of doubt that this book of Bees will in its infancy lie hidden in obscurity…without friends or acquaintance…' Preface to the Reader, 1623 edition of the *Feminine Monarchie*

4   Bloxam, John *Register of the Presidents, Fellows, Demies, Instructors in Grammar and in Music, Chaplains, Clerks, Choristers, and other Members of St. Mary Magdalen College, Oxford*, 1853, Vol 1, p 20 who says that Butler matriculated at Magdalen College, Oxford on 24th November 1581, aged 10, supporting Eva Crane's supposition that he was born in 1571, not 1560, as is sometimes suggested.

5   Hill, Christopher, *England's Turning Point, Essays on 17th century English history*, 1998, Chapter 23, Bookmarks Publications Ltd

6   Ward, W.R ed. *Parson and Parish in Eighteenth Century Hampshire* 1995 Hampshire County Council, p 153

7   Wood, Anthony, Athenae Oxonienses, 2 vols. (1691–2); 2nd edn (1721); new edn, ed. P. Bliss, 4 vols. (1813–20); repr. (1967) and (1969)

8   *A Dictionary of the English Language*, 1768

9   Fuller, Thomas, *The Worthies of England*, published posthumously 1662;

10 Aubrey, John, *Brief Lives*, quoted in *Music at Midnight*, John Drury, 2013, Allen Lane

11 Fraser, H Malcolm, History of Beekeeping in Britain, 1958 reprinted Northern Bee Books, 2010

12 Fraser, H Malcolm, History of Beekeeping in Britain, 1958 reprinted Northern Bee Books, 2010 p 30

13 James 1 was reportedly shown one of the first compound microscopes in 1619, after its invention in Middleburg around 1590. H Malcolm Fraser, unpublished *'History of Bees and Beekeeping'* Archives of the Museum of English Rural Life.

14 Elizabeth Crachiolo, unpublished paper: *Queen Bees, Queen Bess, and the Gender Politics of Butler's Feminine Monarchie*. www.academia.edu] academia.edu/11803648/Queen_Bees_Queen_Bess_a

15 Carlton, Charles *Archbishop William Laud*, 1987, Routledge & Kegan Paul

16 Bates, A.S. *Charles Butler, Vicar of Wootton, 1600-1647* Church pamphlet c. 1952

With thanks to

Ben Taylor, Archivist's Assistant, Magdalen College, Oxford;
Adam Lines, Museum of English Rural Life Archives, University of Reading;
Robert J Hawker;
Paul Hurley of High Wycombe Beekeeping Association;
The Revd Jeremy Vaughan, Rector of the Benefice of Oakley with Wootton, and
Daphne Oliver-Bellasis of Wootton;
Jo Dunbar of Botanica Medica;
Paul Warren & Graham Gardner of Peter Symonds Sixth Form College,
Winchester;
Weald & Downland Open Air Museum, Singleton, West Sussex;
Jeremy Burbidge of Northern Bee Books.
I have enjoyed attending seminars of the Centre for Studies in Rural Ministry,
held at Gladstone's Library, Flintshire and am also grateful to the parishioners
of my own rural parishes in Hampshire for their knowledge and skills in rural
matters.
Particular thanks to Jane, my wife, for her help with editing seventeenth century
English, and for patiently enduring a great deal of talk about the Reverend
Charles Butler over the past few years.

John Owen, August 2016
Steep, Hampshire

Contents

My Book of Bees I divide into ten chapters.

Chapter 1
Of the nature and properties of bees, and of their queen

Among all the creatures which our bountiful God has made for the use and service of man, in respect of great profit with small cost, of their ubiquity or being in all countries, and of their continual labour and agreeable order, the bees are most to be admired.

For first with the provision of a hive and some little care and attendance, which need be no hindrance to other business, but rather a delightful recreation amid the same; they bring in store of sweet delicacies, most wholesome both for meat and medicine… as they well know, who know the rare virtues of honey and wax: a taste of which I will give you in the last chapter.

Secondly, whereas… some country yields one fruit, some another; some bears one grain, some another; some breeds one kind of cattle, some another; there is no ground (of which nature so ever it be, whether it be hot or cold, wet or dry, hill or dale, woodland or common land, meadow, pasture, or arable: in a word, whether it be fertile or barren) which yields not matter for the bee to work upon.

And thirdly, in their labour and order at home and abroad they are so admirable, that they may be a pattern unto men, both of the one and of the other. For unless they are hindered by weather, weakness, or want of matter to work on, their labour never ceases… And for their order, it is such, that they may well be said to have a Commonwealth, since all that they do is in common, without any private respect… They work for all, they watch for all, they fight for all. In their private quarrels, when they are from the hive or common treasury, however you use them, they will not resist, if by any means they can get away… Their dwelling and diet are common to all alike: they have all alike common care both of their wealth and young ones… And all this under the government of one monarch, of whom above all things they have a principal care and respect, loving, reverencing, and obeying her in all things…

If she goes forth to solace herself (as sometime she will) many of them attend upon her, guarding her person before and behind; they which come forth before her ever now and then returning, and looking back, and making withal an extraordinary noise, as if they spoke the language of the knight Marshall's men, and so away they fly together, and soon in like manner they attend her back again.

This I may say because I have seen it[1] although the philosopher is of another mind… If by her voice she bids them go, they swarm, if being abroad she dislikes the weather, or lighting a place, they quickly return home; while she cheers them to battle they fight.

While she is well, they are cheerful about their work; if she droops and dies, they will never after enjoy their home, but either languish there until they be dead too, or yielding to the robbers, fly away with them…

But if they have many princes, as when two fly away with one swarm, or when two swarms are hived together, they strike one of them presently, and sometimes they bring her down that evening to the mantle, where you may find her covered with a little heap of bees, otherwise the next day they carry her forth either dead or deadly wounded. Concerning which matter, I will here relate one memorable experiment. Two swarms being put together, the bees on both sides, as their manner is, made a murmuring noise, as being discontented with the sudden congress of strangers, but knowing well that the more the merrier, the safer, the warmer, yes, and the better provided, they were quickly made friends. And having agreed which queen should reign, and which should die, three or four bees brought one of them down between them, pulling and hauling her as if they were leading her to execution, which I by chance perceiving, got hold of her by the wings, and with much ado took her from them. After a while (to see what would come of it) I put her into the hive again: no sooner was she among them, but the tumult began afresh, greater than before, and soon they fell together by the ears, fiercely fighting and killing one another, for the space of more than an hour together, and by no means would cease, until the poor condemned queen was brought forth slain and laid before the door. Which done, the strife immediately ended, and the bees agreed well together.

Sometimes when one swarm is put to another, though they do not fight, yet will they not agree of their choice in two or three days, keeping their queens close on both sides. But then all this while they are never at quiet day nor night, nor once offer to work, until one of them, being deposed, they are united in the other…

Likewise if the old queen brings forth many princes (as she may have six or seven, yes, sometimes half a score or more, which the surplus of nature affords for more security, in case some miscarry) then, lest the multitude of rulers should distract the unstable Commons into factions, within two days after the last swarm, yes, sometimes (when unkind weather keeps it in overlong) even before it comes forth, you shall find the superfluous princes dead before the hive: I

---

1   The careful observation of his bees' behaviour provides the tone of authority on which Butler's manual rests, but he remains in conversation with classical writers on bee-keeping throughout the Feminine Monarchie.

have taken eight of them up together brought out of one hive, when two were already gone forth with their swarms. For the bees abhor government by the many, as well as anarchy, God having showed in them to men, an express pattern of a perfect monarchy, the most natural and absolute form of government...

The queen is a fair and stately bee, differing from the common ones both in shape and colour: her back is all over of a brighter brown, her belly even from the top of her fangs, to the tip of her train, is of a sober yellow, somewhat deeper than the richest gold. She is longer than a honey-bee, but not so big as a drone, although somewhat longer; her head proportionate but it is more round than the little bees, by reason her fangs are shorter; her tongue not half so long as theirs, for whereas they gather with the one nectar, with the other pollen, she has no need to use either, being maintained as other princes by the labour of her subjects. Her wings are of the same size as a small bee, and therefore in respect of her long body, they seem very short, resembling rather a cloak than a gown, for they reach but to the middle of her train or lower part. Her legs are proportionate, and of the colour of her belly, but her two hind-legs are more yellow; her lower part is long, and half so long as her upper part, more pointed than a small bee's, having in it four joints or partitions, and in each joint a golden bar, instead of those three whitish rings which other bees have at their three partitions. The spear she has is small, and not half so long as the other bees, which, like a king's sword, is borne rather for show and authority, than for any other use. For it belongs to her subjects as well to fight for her, as to provide for her...

Besides their sovereign, the bees have also subordinate governors and leaders, resembling captains and colonels of soldiers; by way of difference from the rest they bear for their crest a tuft or tassel in some coloured yellow, in some mulberry coloured, in the manner of a plume, of which some turn downwards like an ostrich-feather, and others stand up like a hearn-top. And of both sorts some are greater and some lesser, as if there were degrees of those dignities among them. In all other respects they are similar to the vulgar... In less than a quarter of an hour you may see three or four of them come forth from a good stall; but chiefly in Gemini, before their continual labour has worn these ornaments... All of which Pliny... seriously considers that all must with admiration acknowledge that singular wisdom, order and government in them, which in no other creature, man only excepted, (if yet to be excepted) is to be found... Aristotle describes two sorts of bees, the one (which is best) is short, of different colours, and round; the other is long, similar to wasps... And in another place he puts a difference between wild and tame... But these differences my experience has not found: neither do I see how they can be, seeing that swarms of tame bees often fly into trees, and become wild; and the swarms of wild bees are often found, and put into hives. Indeed the wild bees are more angry than the tame, but that is because they are less used to the company of men. Moreover, there is some difference in the size of bees, for those that are laden seem greater and longer than those that are not; the nymphs also, when they first come abroad, are not grown to their full size which afterwards they have, and the old ones wither, and become little again. Similarly, in these three ages their colours also vary. In their middle age they are brown, whereas earlier they are pale, and at the last they turn whitish again. But there are differences of bees in the same stall, and from one stall to another, since these different sorts are in every stall.

The several parts of a bee have their several uses.

Her horns grow in the middle of her forehead, with two joints, one close to the head, the other towards the middle, so that she can put them   forth at full length when she will, and draw them in again close to her head: these are the proper organs of the sense of feeling, by which, with the least touch, the bee suddenly senses any tangible object and therefore they serve to give warning in the dark, and when she is busy, of any obvious thing, living or dead, that might offend her.

Her two cheeks being transparent, like a lantern, serve instead of eyes, though they are immovable, and through which the appearance of visible things are conveyed to the common sense.

For gathering her provision, she has two instruments, her fangs and her tongue; her fangs in the shape of a pair of pincers do not hang like the jaws of other things, one over another, but sideways one against the other, as is most convenient for her uses.

Her tongue is of such length, that her mouth cannot hold it, but being doubled between her fangs under her chin, it reaches to the neck. It is divided into three parts, of which the two outermost serve as a case to cover the third, which being the chief one, the bee in her work puts forth beyond the other, and draws in again as she will. And this third part is likewise parted into three, so that there are five in all.

To set these instruments to work, nature has furnished her with four wings, which swifter than the east wind carry her into all the four coasts of the world, and thence with her precious load bear her back again, until her incessant labour has worn them out.

Her rough and dew-clawed feet apt to take hold at the first touch and are in number six, that she may stand fast on four, while she uses the other two to wipe her eyes, her wings, her tongue, or any other part, and to convey the gathering of her fangs to her thighs.

For her defence she is doubly weaponed. Her fangs she uses when she is not very angry, against all insects, such as other bees, drones and wasps, pinching and holding them often by the legs or wings, and sometimes by the horns, but this is rather a chiding, than a fighting, and a warning, rather than a punishment - though sometimes she bends her spear against them, as if she would kill and slay.

Her spear she is very loathe to use, if by other means she can shift her enemy, as knowing how dangerous it is to herself, for if she chances thereby to strike any hard part, such as the breast or shoulder, she is forced to leave her spear behind her, and so she kills and is killed with the same stroke. Yet when the bees are very angry, as namely when they are assaulted with a multitude of robbers at once or when in the spring a hungry stall forsaking its own home, presses into their hive, they fall suddenly upon them with their poisoned spears... but then they make short work. For by that time they have put up their weapons, and some die immediately, others, losing the use of their wings tumble on the ground like mad things, until in a while they lose their lives too; others when they are wounded, run away in great haste (as having their errand) either drawing on the ground one or more of their legs, or doubling their lower parts toward the ground, or turning the same away to the one side or the other but as many as are stricken, within an hour after will not be able to stir out of the place, and within two or three at the most they will be quite dead. I have looked on[2], while thus they quickly cut off a whole stall, and among the rest, making then no difference, they spared not the queen herself. In the same manner they deal with the drones at the time of the year, when they will not otherwise be beaten away.

But their spears or sting they use chiefly against things of other sorts, as men, beasts, and fowls: which have outwardly some offensive excrement, as hair or feathers, the touch of which provokes them to sting, although such stinging is always mortal to themselves... For the skin having received the sting, holds it so fast, that when they would be gone, they leave both it and part of their entrails which are fastened to it... If they light upon poultry, although their desire is to remain alive, if they come too quickly, they will put forth their spears as soon as they touch the feathers and if by chance they hit a hard part, the sting sticks fast, as in the skin, and therefore goose-wings are not to be used in the hiving of bees.

Likewise, if they light upon the hair of your head or beard, (apart from when they come home laden, or the weather is cold) they will sting, if they can reach the skin, although wool and woollen do not offend them; and if being otherwise made angry, they strike their spears in woollen, they can easily pull them out again. But the nap of new [3] fustian displeases them, because it seems hairy, and the stuff is so soft, that it holds the sting. As a result of which, such apparel is not suitable among

2   'I have looked on': Butler writes as a naturalist, basing his views on the evidence before him, in contrast
     to some of the more fanciful notions about bees found in the classical texts like Virgil in Georgics: Book
     IV.
3   Cloth made of cotton and flax

bees, as also the leather in gloves or otherwise, for as soon as they touch it, they will strike, if they are to the smallest extent moved, and their spears they cannot recover again. Velvet in the facing of hats or elsewhere, angers them as much as anything, making them strike as soon as they touch it, but it has no power to hold their spear.

When they are angry, their aim is most commonly at the head, and chiefly about the eyes, as if knowing that there they may do most harm, for that part swells most and longest, and yet I never heard that any ever stung the very eye, as if they were forbidden to touch that tender part. But the bare hand that is not very hairy, they will seldom or never sting, unless they are much offended.

When you are stung, or any in the company[4], though a bee has struck but your clothes, especially in hot weather, you had best be packing as fast as you can, for the other bees, smelling the rank savour of the poison cast out with the sting, will come about you as thick as hail, so that fitly and truely did he express the multitude and fierceness of his enemies, who said, *They came about me like Bees*[5]. Then there is no way to appease them but flight; the more you resist, the fiercer they are. They are like incorrigible shrews; there is no dealing with them but by patience, though when they sting they are sure to have the worst. For the wound endangers neither life nor limb: two nights sleep will take away the swelling, and two minutes the pain, (unless it is in very rheumatic or humorous bodies, of which sort I have known some so swollen and disfigured with that little stroke, that you would scarcely know them by their savour in five or six days after.) But on the other side, whereas the wasp, hornet, and bumblebee often sting without any hurt to themselves, the bee only stings once, and she leaves her spear and entrails, more or less behind her… For within four and twenty hours after, or, if much of her entrails come forth with the sting, within half that time she dies. But the spear retaining life when the bee is gone, if it is not immediately pulled out, it will work itself into the flesh up to the hard end, and so cause the pain and swelling to be both greater and longer. Therefore when you are stung, instantly wipe off the bee, sting and all, and wash the place with your spittle: so shall you prevent both pain and swelling, which otherwise nothing but time can cure, for the poison is so subtle, that it quickly pierces the flesh, and the wound so little, that no antidote can follow after. Yet I have heard commended for a remedy, the juice of houseleek, of rue, of mallows, of yew, of a marigold leaf, of hollyhock and vinegar, of salt and vinegar, and many other things…

But if you will have the favour of your bees that they sting you not, you must avoid such things as offend them: you must not be (1) unchaste or (2) unclean:

---

4   Butler had help in managing his bees, and would have needed it, given that he kept sufficient hives to raise a dowry from honey sales for his daughter, Elizabeth, at her marriage in 1633. His wife, who died in 1628, was likely to have been his main helper.

5   Psalm 118 verse 12

for impurity and sluttiness (themselves being most chaste and neat) they utterly abhor; you must not come among them (3) smelling of sweat, or having a stinking breath, caused either through eating leeks, onions, garlic, and the like; or by other means; the offensiveness of which is which is corrected with a cup of beer. Therefore it is not good to come among them before you have drunk; you must not be given to (4) surfeiting and drunkenness; you must not come (5) puffing and blowing them, neither hastily stir among them, nor violently which not only increases their anger, (especially in hot weather...) but incites others to take their part, and if by striving and striking you chance to kill one, the bees soon perceive it by the strong smell of the humour[6] (for she smells them as if she had stung) and will be eager upon revenge, so that by no means can they be pacified, until they have the field.

Defend yourself when they seem to threaten you, but softly moving your hand before your face, gently put them by, and lastly you must be (6) no stranger to them. In a word, you must be chaste, clean, sweet, sober[7], quiet, and familiar: so will they love you, and know you from all others.

At any time, when nothing has angered them, one may boldly walk along by them, but he who stands still before them within the space of a [8]perch in the heat of the day, it is a marvel but one or other spying him from the hive, will have a cast at him.

If you have anything to do about your hives, the most suitable time is in the morning, when the bees are newly gone abroad, and in the evening before they have come in, for then the weather being cool, and the company few at home, they are not so apt to be quarrelling, unless they are much provoked. Likewise at other times of the day, when the weather is cold, wet or windy, they are patient enough.

But about noon in hot weather, and especially when they have tasted of the honey-dews, they are soon angry, and very eager.

But whenever you have occasion to trouble their patience, or to come among them being troubled, it is better to stand upon your guard, than to trust to their gentleness. For the safeguard of your face (which they have most mind to) provide a purse-hood made of coarse boulter cloth, to be drawn and knit about your collar, which, for more safety, is to be lined against the eminent parts with woollen-cloth. First, cut a piece about an inch and a half broad, and half a yard long, to reach round by the temples and fore-head from one ear to the other, which being sewn in its place, join to it two short pieces of the same breadth

---

6   Butler accurately observes what we now understand to be the release of pheronomes in a hive, as a
      means of communicating danger or distress to the rest of the bees

7   Qualities also valued in beekeepers by Virgil and other classical writers

8   About five and a half yards

under the eyes, for the balls of the cheeks, and then set another piece about the breadth of a shilling against the top of the nose. Instead of this, you may use a cypress band or a boulter, having a handkerchief between your forehead and it, to bear it out from the skin, and your hat on your head to hold it fast. And if they are so earnest that you fear them stinging your hands, put on a pair of woollen cuffs or gloves. When you have on this helmet and gauntlets, as a man armed at all points, you may boldly deal with them, being out of the danger of their poisoned spears. At other times when they are not angry, a little piece half a quarter broad to cover the eyes and parts about them may serve, for then, though it is in the heat of the day, unless they may strike above the eyes, they care not to strike at all.

Towards cattle, which have not the reason by flight or otherwise to save themselves, bees are more dangerous. A horse in the heat of the day was looking over a hedge, on the other side of which was a stall of bees, and while it stood nodding with its head, as its manner is, because of the flies, the bees fell upon it and killed it. Likewise, I heard of a team of horses stretching against a hedge, which overthrew a stall on the other side, and so two of the horses were stung to death. I doubt not but that through negligence many such mischances have happened elsewhere. For this thing has been long since observed by that great philosopher Aristotle…

And such are the sorts of bees, with their integral parts. Among which, though there do not appear those outward organs of scent which other animals have, nor is seen in the head that inward principal part, which is the fountain and seat of all senses, imagination, and memory, yet have they the senses themselves, both outward and inward: which their subtle and active spirits do excite and quicken, for the works of their curious art and singular virtues…

Of all the five senses their sight seems to be weakest, and weaker too when they come home loaded, than when they are unladen and being loaded they are weaker on foot, than when they are flying. If, when they come home loaded, they alight beside the door, they will go up and down seeking for it, as if they were in the dark, and unless by chance they hit upon it, they must fly again before they can find it. As many as fall beside the stool when it gets dark, ten to one they lie abroad all night, yes, if at such time being troubled by anything they come forth from the stool, though then they are fresh and strong, and will leap up and down, run and fly to and fro, until they become weary, but by no means can they find the way in again. And therefore it is that when they fly abroad, they take such pains at the door in rubbing and wiping their glazed eyes, that they may the better discern their way forth and back.

But their sense of smell is excellent, by which when they fly aloft in the air, they will quickly perceive anything under them that they like, such as honey, resin, or tar, even though it is covered. As soon as the honey-dew is fallen, they immediately change direction, even though the oaks which receive it are afar off, which the poet[9], speaking of the excellence of some creatures in this sense before others, expresses in this way…

And by this sense they find out any strange bee, which is not otherwise to be known from their own company, and that in the dark hive, where, when they are disposed, they will by the same means cull the drones, yes, and pull out the young drones[10] that are shut up in the cells, not meddling with any of their own sex.

Their hearing and feeling are very alert. If you touch their hive but lightly, or the stool, or the ground near it, they immediately perceiving it, make a general noise, although Aristotle[11] doubts whether they hear… But if they did not hear, to what purpose is that music made in the hives, before the swarming…?

And of their sixth sense I make no question, since they are used to things of many different tastes, although there may seem the less use of it, because their smelling is so perfect.

And such are their outward senses. The inward qualities of their minds are far more excellent. Their curious art and workmanship is to be admired rather than imitated of men.

Their unique virtues are no less admirable.

In valour and magnanimity they surpass all creatures: there is nothing so huge and mighty that they fear to set upon, and when they have once begun, they are invincible, for nothing can make them yield but death, such great hearts do they carry in such little bodies. In private wrongs and injuries done to their persons (for which cause men will soon quarrel) they are very patient, but in defence of their Prince and Commonwealth they do most readily enter the field…

By which means appears their unique fortitude, no less than their prudence does in the government of their Commonwealth, beside which, their wisdom and knowledge in other matters is very great, in regard to their enemies, their fellows and friends, the drones (when they have too many, and when they need them not at all), also the times and seasons of the year. Their wit and dexterity, as well in gathering as in working their sweets, is inimitable.

Moreover, as skilful astronomers, they have fore-knowledge of the weather… And in stormy and windy weather, it is a wonder to see what cunning those that are abroad use to shift the wind when they come home laden: how they fly low by the ground, among the bushes, in the lanes and lee-sides of the hedges…

---

9   Lucretius, c. 99 – 55 BCE, Roman poet and philosopher.
10   *Cephens*
11   Aristotle, 'The History of Animals' approx. 350 BCE

But above all, one excellent skill they have, which the most excellent females, however much they desire it, must yield themselves to want, for they know certainly when they breed a male, and when a female, which appears in this way, that they lay their drone-seeds in a wide comb by themselves, and the nymph-seeds in the rest, which are of a smaller size. So that near the top of a most fine box, in which the Host was laid, the choir of bees are singing around it, and keeping watch in the night, as monks do in their cloisters. The bishop therefore, taking the host, carried it with the greatest honour into the church to which many resorting, were cured of innumerable diseases[12].

I don't doubt that some incredulous people will quarrel with this story as well as with the former, and ask questions of it, since the combs in the top of the hive, which are no more than half an inch apart, one from another, how could there be room for a box of that breadth that would contain the Host? And then being there, how might it be seen by the bishop, seeing those spaces are always filled with bees, and the story says, that they were then singing about it. Therefore, perhaps, they will suspect the whole narrative, supposing it rather to be an unadvised device of some idle monk, which, if he had consulted with those who have skill among bees, might have made his tale more probable. Alleging moreover, that therefore there is no mention made of any particular person, time, or place, lest the circumstances should disprove the matter itself. All of these objections I could as easily answer as the former, if I thought it necessary. But now because some may be as ready to mistrust my narration, as others are to object against the truth of the stories, I will hear on my own behalf for their satisfaction, and set them down in my author's own words…[13]

In that story we may note, that besides the wonderful knowledge and devotion of the bees, there is an incredible power and virtue also. For this God whom they kept and compassed, is said to have the gift of healing, which others, though of as good a making, we know do lack. The conclusion, which my author necessarily infers on this point, is better than all the rest… But if you will grant me that close by is proved the incredible knowledge and skill of the bees, for my part I will urge you no farther.

---

12 This passage is based on an elaborate story by the sixteenth century priest, Thomas Bozius in 'De Signis Ecclesiae' from whom Butler quotes extensively. It at first seems out of character with the analytic and rational tone of FM. The 1623 edition of the Feminine Monarchie reduces the more extensive passage found in the first edition of 1609. By 1623, the practice of carrying the sacred bread into church, the Host, representing the Body of Christ, was officially outlawed in the Church of England, and this may be why Butler edits it down for the second edition. It shows how much Butler was in touch with Catholic teaching and thought, and that he remained in sympathy with Archbishop William Laud's teachings. Elizabeth Crachiolo has extensively explored the political nature of this passage in an unpublished paper: *Queen Bees, Queen Bess, and the Gender Politics of Butler's Feminine Monarchie*. www.academia.edu] academia.edu/11803648/Queen_Bees_Queen_Bess_a

13 A long quotation in Latin follows, from Thomas Bozius, *De Signis Ecclesiae*, Book 14.

In the pleasures of their life, the bees are so moderate, that perfect temperance seems only to rest in them.

Also, in their own Commonwealth, they are most just, and not the least wrong or injury is offered among them. But indeed I cannot much commend their justice towards strangers, for all that they can catch is their own, unless they may be excused in this respect, that the bees of different hives are at deadly feud, or rather as Kingdoms, that are at defiance one with another.

Their chastity is to be admired…

They are born not as other living creatures; they allow their drones among them only for a season, by whose masculine virtue they strangely conceive [14] and breed for the preservation of their sweet kind. Which strange kind of breeding the Philosopher says is apparent to sense and reason…

For cleanliness and neatness, they may be a mirror to the finest dames. … For neither will they allow any fluttery within, if they may go abroad. … neither can they endure any unsavouriness outside and close to them… And as for their persons (which are lovely brown) though they are not long about it, yet are they curious in trimming and smoothing them from top to toe, like sober matrons, who love to go neatly as well as plainly. Pied and garish colours belong to the wasp, which is good for nothing but to spend and waste.

About the age of bees there are different opinions: some think they may live four or five years, yes, some six or seven; Aristotle speaks of a longer time … These opinions are grounded on this, that they see a stall[15] sometimes continue so long, before the bees die altogether. But this continuance is only by succession, and so might they live in their fecal matter, were it not for the rottenness of their combs, the hardness of their home, and the offensive smells in abundance which don't allow them to abide there… But the truth is, a bee is but a year's bird, with some advantage.

For the bees of the former year, which until Gemini in the next year look so youthful, that you cannot discern them from their full grown nymphs, which have been bred that spring; they do from that time change with obvious difference, for the young bees continue to become great, full, smooth, brown, and well-winged; the old grow little, withered, rough, whitish, ragged-winged and in addition so feeble, that when they come laden home, if anything stands in their way, yes many times, though there be nothing, they fall down, and being laden cannot rise again, and then either a little cold or wet in the day, or the night's dew kills them. You may daily find, especially in Cancer and Leo, some dead, and some half-dead before the hives, and some alive and vigorous, which yet can never rise again. Some of them will hold out so long, until their wings are more than half worn, but

---

14 The actual process by which the queen is fertilized in a drone cloud is not known to Butler
15 Honey-getting colony and/or hive.

by Libre you shall scarcely see one of them left.

The young bees, as best able, bear the greatest burdens, for they not only work abroad, but also watch and ward at home both early and late. When the need is there, they hazard their lives in defence of the rest; they beat away the drones, and fight with other bees and wasps, and assault with their spears whatever else offends them. They carry their dead out to be buried, and perform all other offices. But the labour of the old ones is only in gathering, which they will never give over, while their wings can bear them, and then when they cease to work, they will cease also to eat, such enemies are they to idleness. And therefore generally they die in their delightful labour, either in the field or coming home... Sometimes as well in summer as winter the bees take pleasure to play abroad before the hive, especially those that are in good condition, flying in and out, and about, so thickly and so earnestly, as if they were swarming or fighting, when indeed it is only to solace themselves: and this chiefly in warm weather, after they have long been kept in...

The bee is by nature very tender, soon chilled and killed with cold, which the bumblebee, the wasp, the moth, the gnat, and other little flies can endure, and most of all then, when by reason of long restraint, their bellies are overfull. The first that fails in them, when the cold begins to prevail, is their wings, so that they cannot rise to their hives to help themselves by the heat of their fellows. How to recover them, even when they are quite dead, is referred to in my notes.

The Bee therefore excelling in many qualities, and it is suitably said in the Proverbs,

	{Profitable	}	
	{Laborious	}	
	{Loyal	}	
	{Swift	}	
	{Nimble	}	
	{Quick of feet	}	
As	{Bold	} as a Bee	
	{Cunning	}	
	{Chaste	}	
	{Neat	}	
	{Brown	}	
	{Chilly	}	

These wonderful parts and properties of this little creature, what are they but so many evident proofs of the infinite power and wisdom of the Creator?

Chapter 2
Of the bee-garden and seats for the hives

For your bee-garden, first choose some plot near your home, that the bees may be in sight and hearing; because of swarming, fighting or other sudden happening, during which they may need your immediate help. While the stalls are few, your garden of herbs and flowers will serve... But when they are grown to a sufficient number, they require a square green plot fitted for the purpose.

2. See it is safe, and securely fenced, not only from all cattle, (which if they break in, may quickly spoil both the bees and themselves) and especially from swine (which by rubbing against the hives, and tearing the hackles aggressively, are most likely to overthrow the stalls); but also from the violence of the winds, that when the bees come laden and weary home, they may settle quietly.

The north fence of your garden should be close and high, that the cold wind of the coast, (which blowing against the bees coming home weary, would throw down and kill many) may be altogether kept from them. And therefore, if it may be, set your bees on the south side of your house.

The east fence also should be good and high to keep from the bees as well the sun, as the wind. For the sun rising does often bring them forth, when the air is colder than they can endure; and the east-wind being cold and sharp is very unkind for bees, especially in the spring.

But in no way let the place be shadowed from the south sun, for that does not only dry the hives and relieve the bees in the winter and spring, but also causes them to swarm in summer[16], if it is not extremely hot and dry.

Nor yet from the sun's setting, because in calm and pleasant weather the bees will be in the field after the sun is down, even as long as they can there see, and if when they return, they find it dark at home, many of them, their sight being but dim, fall short or wide, and flying and running to and fro until they become weary, at length they yield to the cold dew.

Otherwise let the fences be as against the south and west-winds also, as may be, for although they are not so cold and bitter as the others, yet are they no less violent, and more frequent, so that they also do much harm, especially in the spring. And therefore if at that time of the year, in rough and boisterous winds, you find that the garden fences do not sufficiently guard and defend them, then is it good to set up waxed or lined hurdles, or some other screen between them

---

16   Swarming is the only means of increasing bee stocks, and is therefore essential for the seventeenth
      century beekeeper

and the weather. For though they can shift abroad in the strongest winds, as a ship that has sea-room, yet are they easily overthrown at the hive, as a ship is soon wrecked at the haven.

A house or wall is best for the north fence, and a quick-set hedge for any of the other three: it may serve also for the first, especially if it is thick.

3. That the place be sweet, not annoyed with any stinking savour. I have known a stall in the spring, being sufficiently provided of honey, and having bred young, to forsake all, because of poultry that roosted in a tree over them...And yet the smell of urine does not offend them: no, they will be very busy where it is shed. It is thought they use it for medicine...

4. Ensure it is neither very cold in winter, nor very hot in summer... A simple shower is ruinous in both seasons, because in winter it is over cold, and by that means quickly chills the bees that a light upon it, and in summer it causes them to lie forth through excessive heat. A grassy ground therefore is best at all times, but let it be kept not[17] in summer, and dry in winter, for long grass and weeds about the hive harbour the bees' enemies and hinder both their passage in and out, and their rising again when they fall short, and water if it stands, as it will be often to yourself, so is it dangerous to your bees for chilling and drowning them. And as the parts about the hives are to be kept not and bare; so are other places also, which the swarms use to play and pitch, whether within or without the garden, freed likewise from long grass and weeds, and more so from beans, peas, hemp, and such high things, for the young weak nymphs falling in those shady places - unless the weather is warm and dry - are in danger of being chilled before they can rise again. For which reason the swarms do usually refuse to stay and settle about such places, and then if windy or cloudy weather forces them not to go further, they must either go home, or alight upon some other hives, where, without your patient skill and diligence, they are likely to be all lost.

5. That it is conveniently beset with trees and bushes fit to receive the swarms, such as plum-trees, cherry-trees, apple-trees, filberts[18], hazel trees, thorns etc. Which they will the more delight to a light upon, if convenient boughs are hanging out alone from the bodies, the twigs below standing in their way are pruned, and the weeds and grass underneath are cut away close to the ground. Although, if they are willing to stay, they will not refuse a dead hedge, a lavender border, or similar, or sometimes the bare ground. For want of trees, some have stuck up green boughs, and the bees have lighted upon them.

The place being thus fitted, the seats are to be provided: which, whether they are stools[19] or benches, must be set a little shelving, that the rain may neither run

---

17   Presumably the grass should not be kept long for the purpose of hay making
18   Cultivated hazel-trees
19   The stand on which the hive is placed

into the hive, nor stay at the door.

To see many stalls upon a bench (as many are used to do) is not good: for in summer it may cause the bees to fight, as having easy access on foot to each other, and standing so near, that they shall sometimes mistake the next hive for their own, and in winter the bench will be always wet, which loosens the comb, rots the bottom of the hive, and offends the bees: and the mouse at all times has free passage from one to the other, without fear.

The single stools therefore are best. And yet it is not amiss to set most of your swarms upon benches, about the old stalls, from whence remove them to the stools, when the stalls are taken, and then set up the benches until another year. Yet I prefer single stools set two feet apart, though they are laid flat on the ground, but it is better to rear them with four legs, though little and short. If they are twelve or thirteen inches, three or four inches may be forced into the ground for their surer standing.

The best stools are of wood; those of stone are too hot in hot weather, and (which is worse) too cold in cold.

For their size, they should not be above half an inch or an inch outside the hive, save only before, which needs the space of three or four inches, that the bees have room enough to a light upon, especially then, when the sight of a rainy cloud sends them thronging home. Which fore-part from one side to the other, is to be cut shelving that it may the better avoid the rain. And therefore if the hive is fifteen inches across, the stool should not be above fifteen or seventeen inches one way, and nineteen or twenty at the most the other way.

These stools should be set towards the south, or rather a point or two into the west, that the hive may somewhat break the east-wind from the door, and that the door may be lightened by the sun's setting, when the bees return late and laden from the fields, and therefore it is to be wished that the garden fences stand accordingly.

They should stand in straight ranks or rows from east to west, five feet from one another (measuring from door to door) and from north to south, six feet from one another.

Likewise let them stand as far apart from three of the fences, as they do from one another. And so a plot of fifty feet square, will receive seven ranks of nine stools a piece, with the space of eight feet before them, which if it were bigger, were so much the better.

For want of room or stools, or understanding, many bee-masters do set their stalls nearer together. But the greater distance is much better, not only that you may have room enough to go round about every one, to see and mend what is amiss, but also that the bees, when they come home in haste, especially when a swarm goes back again, may be sure to fly into their own hive. For if they stand near together, at such time many will take the next hive for their own, and then they fall into close contest, and the nymphs, when they go first abroad, will on that occasion the sooner mistake their hive, which if they do, they die.

The manner of placing the stools in your garden, with the distance of the ranks, I have here expressed.

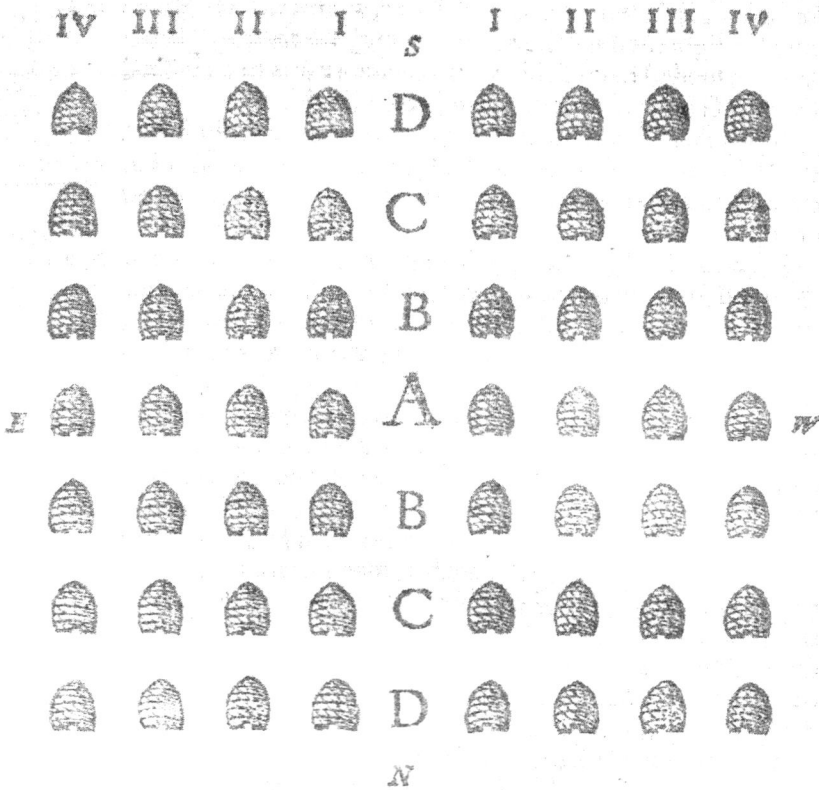

This critical number of nine times seven, is an adequate or rather complete store for any one garden, though large and alone, which being well ordered, will yield the bee-master the better part of a liberal maintenance, if any be so happy to attain to it. So that I see no evil at all in this number, although the sixty three years of man's age, being likewise called critical (because it arises from nine sets of seven, like so many decisive points or ladder-rounds) be counted of some, and those no small fools too, a parlous and ominous time, more dangerous for death, than all the other years of their life. For which notion if you see no reason, note that it is grounded upon good observations, for this is certain, that a ladder of nine rounds[20] has been fatal to many...

Answerable unto this critical squadron it is proper you have at hand a register, containing the several ages and yearly increases of all your stalls[21]. By these entries you may be directed every year, to those stalls which are to be taken and those which are to be kept for store, which is the chief point of a thriving bee-master. This register may be a synopsis or table drawn upon a sheet, or half sheet of paper, divided into sixty three squares, or as many as are necessary for the stalls in your garden, having first the four coasts, E.S.W and N, noted in the outsides: secondly, the middle row of squares from S to N distinguished by letters, the first square being marked about with D, the second with C, and the third with B, which are southern; the fourth (being the chief one, mostly in the middle, to which all the squares in the table have reference) with A, the fifth with B, the sixth with C, the seventh with D – the last three are northern - and thirdly, the first row of squares next to the letters on both the E and W side, noted in the top or south-part with one I, the second on both sides with II, the third with III, and the fourth with IV.

The table thus drawn, when you have set a swarm on any stool in the garden, mark in what letters its rank it is, what number from the letter, and whether eastward or westward: and in the square answering to that begin its register, setting down first the two last figures of the year of the Lord, then for a prime swarm, a double circle, for a casting half a double circle, then the day of the month in which it was swarmed, writing M for May, I for Iune, J for July. The next line begin with the next year: if it swarmed set down a crossed circle, and the day of the month if it swarmed again, set down in the same line a half circle, with a down-right stroke, and the day of the month if it did not swarm, but were full to the door, set down a circle with a full point in it; and if it did also lie out, set down a circle with a blotted circle in it. If it did neither lie out nor was full, set down a void circle.

And then do likewise for all the years that this stall endured. When the table waxes full, after the harvest make a new: taking out of the old the register of those that live. By this means you may certainly know the age, and yearly increases of any stall in your garden: and so guess whether it is fitter to kill or to keep.

---

20   This may be a reference to the Masonic Ladder of Nine Rounds
21   Butler's precise observation and use of accurate records in beekeeping makes *The Feminine Monarchie* an early work of English natural history

Also the stools should not stand above two foot from the ground, because of the wind, nor under one foot for the dampness of the ground in winter, which would make the hives moist and musty; and for the heat of the ground in summer, which in hot and dry weather would make the bees lie out, and so hinder both their work and swarming.

The best height is between eighteen and twenty inches. Yet if you have many, it is convenient that the more northward ranks should stand higher, and the more southward lower, descending by degrees from two foot to one. If there are two rows of stools, let the first stand two foot from the ground, the next eighteen inches, and the benches or swarm-stools one foot or less. If there are three ranks beside the benches, let the second be twenty inches, and the third sixteen etc.

This unequal height of ranks may as conveniently be effected, though the stools are all equal, by the unequal levelling of the ground, which in a great bee-fold is best.

The stone-stools must be footed as they may; the fashion of each place where they are used will direct you. But the planks or wooden stools are either to have four feet made of the heart of oak, or of some other lasting wood; or to be fastened to one foot with two wooden pins, which foot let it be made of sound timber five or six inches over, and of that length, that it may be set between fifteen and eighteen inches in the ground.

Chapter 3
Of the hives, and the dressing

In some countries they use straw hives bound with briar; in some twig hives made of privet, willow, or hazel, daubed usually with cow-cloom[22] tempered with gravelly dust, or sand, or ashes.

The straw hives when they are old and loaded, do usually sink on the one side (especially if they become wet) and so break the combs and let out the honey, in which cause, first see that they are hard wrought, and then split them strong with a cop[23] fitted to the top of a hive.

The wicker hives will still be at fault, and lie open (if they are not often repaired) to wasps, robbers, and mice. Any of these, if they find but a little crack, will dig their way in, and the mouse (unless the twigs are closely wrought) even though it finds none.

Both these hives, if they are not well covered, are subject to wet, which makes them musty, and, if it is much, rots the combs, and destroys the bees. But the heat in summer, the cold in winter, and the rain at all times does soon pierce the wicker hives, for which reason it is good to double-daub them.

All things considered, the straw hives are better, especially for small swarms.

The bees best defend themselves from cold, when they hang round together in the manner of a sphere or globe (which the philosophers account the most perfect figure) and therefore the nearer the hive comes to that fashion, the warmer and safer the bees. But of necessity, the bottom must be broad, for the upright and sure standing of the hive, and for the better taking out of the combs, and the top must rise some two or three inches higher than the actual form of a globe, to hold up the hackle[24], and to shun the rain, which yet, where the hives are covered with pans, is not necessary. Otherwise let your hives vary no more from this round figure, then needs most: as where it is within from the top to the skirts seventeen inches, in the middle or widest place through the centre fifteen inches, and at the skirts, thirteen, in this shape:

---

22   A mixture of cow-dung and clay
23   *Top*
24   A separate cover made of straw, to fit over the hive

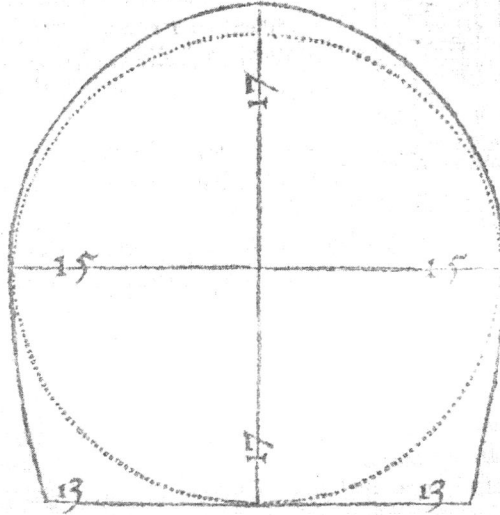

This shape with its dimensions will contain six gallons[25], and the reduction of one inch in each dimension, reduces it by a gallon[26] in the content.

The best that I have seen are wrought by Thomas May of Sunning, about one mile from Reading[27].

Hives are to be made of any size between a bushel[28] and half a bushel, that any swarm, of what quantity or time whatever, may be suitably hived; less than half a bushel will not contain a competent stall, and more than a bushel is found too big for any company to continue, and thrive together.

The middling size of six gallons, or within half a gallon[29], under or over, as fitly containing the natural quantity of a good stall, is most profitable.

Have always hives enough of all sorts (but most of the middling size) in store, lest they are ailing when you should use them.

The best time for making them, whether they are straw or twigs, is in the three still months of winter, Sagittar. Capr and Aquar. for then the straw, briers and twigs are best in season, and then is it best to provide them, because then they are best cheap.

Your hive being ready is thus to be dressed. First, rake away all those conspicuous straws, twigs, and other offensive jagged ends that are fastened in the hive, making the inside as smooth as may be, for these obstacles being many,

25   Three pecks
26   Gawn
27   This mention of *Thomas May of Sunning* suggests that Butler was part of a wider network of beekeepers which shared both ideas and equipment. Sonning-on-Thames, Berkshire is under 30 miles from the parish of Wootton St Lawrence, where Butler was vicar from 1600, until his death
28   A measure of capacity used for corn, fruit, etc containing four pecks or eight gallons.
29   *Pottle*

if they cause not the bees to forsake the hives, yet will they much trouble and hinder them. You may hear them (especially in the night) scraping and gnawing three or four days after they are hived, yes, sometime a week together, as though there were mice in the hive - and in straw hives a long time after.

If you need but few hives you may prune them clean with your knife; if you must use many, then, having wet the skirts with a cloth, singe or sweat the inside, but first and last rub it well with a rubber, which is a piece of rough grindstone or sandstone, as great as your hand can hold.

2. The hive being pruned, put slips of wood[30] in it, three or four, as the size of the hive requires; the upper ends of which set together at the top of the hive, and the bottom ends fasten below in equal distance, about a handful above the skirt. In a wicker hive let the upper ends rest against the middle of the staff, and the bottom ends against the parts of it between the twigs; and in a straw hive, let the upper ends together in a cop, and the bottom ends against the briars or threads, between the third and fourth row.

The cop is a round piece of wood an inch or two thick, whose lower surface is flat, with a hole in the middle half an inch deep, for the spleets to rest in; and the upper is convex, turned or hewed fit to the concavity of the top of the hive.

And for the spleets, take a straight hazel or willow-stick, quarter it if it is big enough, else split it: then shave and smooth the clefts, and having brought them to a convenient strength and length, cut the lower ends forked, to stay against the hive's sides, and the upper ends somewhat picked, and of that size that they may fitly join in the cop or middle of the staff, with their backs leaning hard and fast, one against another.

And this is a handsome, easy and sure way of spleeting: it is also good for drawing the coombs without breaking, and for keeping the hive from sinking and from tearing at the top. Besides which there are different sorts of spleeting, which it is needless to rehearse, for every country has its fashion.

3. Lastly, in swarming time season the hives that you mean to use, rubbing them with sweet herbs such as the bees love, as thyme, savory, marjoram, balm, fennel, hyssop, mallow, bean-tops. And when the swarm is settled, take the hive that you think fit for it in size, and with a branch of hazel, oak, willow, or any of the aforesaid herbs, but chiefly with a sprig of that tree on which the swarm alighted, wipe the hive clean, and then dipping it into meth[31] or fair water mixed with a little honey, or with milk and salt, or, for a need, with salt only, besprinkle the same.

---

30 *Spleets,* used to stop the skep wall collapsing under the weight of honey
31 Metheglen (Meth.) is a drink of spiced mead which originated in Wales

But if the hive has been used before, after you have pared away the wax as clean as may be, if you think the former dressing will not make it sweet enough, then let a hog eat two or three handfuls of malt, or peas, or other corn in the hive: meanwhile so turn the hive, that the foam or froth, which the hog makes in eating, may go all about the hive. And then wipe the hive when the bees are so difficult to deal with, that they will not otherwise abide.

And thus are the hives to be prepared and dressed, before they receive the bees. Now will I show you how they are afterwards to be fitted and furnished.

First, let them always be well covered, that they may be safe in summer from heat, lest the wax melting, the combs fall down; in winter from cold, lest it kill the bees; and at all times from rain, lest it corrupt first the hive, afterward the combs, and at last the bees also.

In some places (where the stalls are not many) they use earthen covers, but these do not defend the lower part, and in summer are too hot.

[32]The best cover for hives is a thick hackle…which is thus to be made. Take four or five handfuls of wheat or rye leafed out of the sheaf, which being bound up severally, beat out the corn; and then casting away their bands, draw out the ears of each handful longer on one side than on the other, and putting the long sides together (so to make the head in the form of a pyramid or sugar-loaf, for shooting away the rain) bind them all in one under the ears, as hard as you can.

The head is to be covered or bound with a cap, of which there are two good fashions, the one wreathed, the other platted.

The wreathed cap is made in this way: having bound the bundles all fast together with a thong, cord or other strong string, leaf out of the sheaf almost a handful of the strongest straw, and lay it in {to} soak about a quarter of an hour. Being thus prepared, take out of that wet bundle a handful[33] of forty or fifty reeds or straws, and laying half of them one way and half the other, that the band may be of equal size, take them up together, and then mingling one end of the bundle with the middle reeds of the head, and twisting them fast together in your hand, allow the band fibre[34] to double in the very top of the head, and so begin to bind the head round, working downward, and still twisting the band as you go. When that handful is well-nigh wrought up, take out of the wet bundle so many more reeds prepared as before, and when you have mingled one end of that with the end of the first bundle, holding them in your hand twist them fast together, and so continue your work, always binding as hard as you can, and bearing up every roll close to its fellow. When you are come down to the string, loose it, and bind

---

32 Many of the words which Butler uses to describe skep bee-keeping are now obsolete cf. *'The text still merits study by the serious beekeeper, although many of his technical terms fell out of use in the nineteenth century.'* (Charles Butler, Oxford Dictionary of National Biography, 2004

33 *Litche* handful, bundle

34 *Harle*, a filament or fibre

the last or lowest roll in the place of that, making fast the end, by forcing it up between the head and the cap with a forked stick and a mallet.

The platted cap is wrought contrary to the wreathed, for whereas that is begun in the crown and wrought downward toward the right hand, and is made fast in the neck, this is begun at the neck, and wrought upward toward the left hand, and is made fast in the crown, after this manner.

First take a handful of strong reeds, and having wetted and wound it a little, put it about the neck of the hackle, and knitting the ends in a half knot, gird the hackle hard with it (your assistant holding one end, while you pull the other), then to make this collar fast, wrap each end about it, forcing them between the collar and the head with the fork and mallet. Otherwise you may make a strong collar of a small width. The collar thus fitted to the neck, set the hackle between your legs, as you sit or stand, with the knot outward, and then, to begin, take up a handful of the ears (about the size of the top of your finger) next to the aforesaid left end of the collar, and laying this end between it and the head, turn the top of the head downward, and so leave it. Then take the next handful and laying the first between it and the head, turn the first downward, and so leave it; then likewise take a third handful, and laying the second between it and the head, turn the second downward, and so leave it, likewise the fourth, and so forth, working thus round, till you come to the crown, and platting still the bundles hard, and close to the head. But when you come to the other end of the collar, take that in for a handful. If the bundles are too short for the work, pluck them up higher about the neck as you go. When you have wrought up to the crown, knitting the four last or top bundles in a true-loves-knot, make all fast.

The hackle thus made of four or five handfuls will contain in compass about the neck, close under the cap, between sixteen and twenty inches; sixteen will serve for the smaller hives, and twenty for the greatest, although they be five feet about.

For the length of the hackles, each one is to be fitted to its hive, so that the skirts of which may reach to the stool, or within half an inch of it round about, save only before, where it must be pared somewhat shorter, that the bees' passage is not hindered.

And then with a small pliant garth or belt of bethwine[35], bramble, briar or the like, gird the hackle close to the hive, lest the wind disorder it. If there is any crook or bending in the belt, put that in front, that the hackle, bearing in that place farther out, may shoot the water from the door. Otherwise, for that purpose, put the belt somewhat higher before, then behind.

---

35 *Bethwine*, a name given locally to various twining plants, including convolvulus

The hackle fitted and placed is now and then to be removed, not only to come across mice, moths, spiders, earwigs etc which harbour under it, and to see what breaches the mouse and titmouse have made, but also to air the moist hive, and this in a warm and windy day after much wet.

Next keep the hives always closed up for the defence of the bees against their enemies. The best cloom for that purpose is made of cow's dung, but to harden it, temper it with lime or ashes, with sand or gravel, which are also good against the gnawing of mice. With this cloom close up the skirts and gaps[36] of your hive, that there is no way into them, but only by the doors.

And being thus safely shut, move them not without urgent occasion, for often lifting up the hive, and letting in the open air does discourage the stall.

But whenever you are occasioned so to do (the bees being stirring) lest any be crushed between the skirts and the stool in setting it down again, lift up one side with a little tile shard, which, when the bees are quiet, take away, and see the hive closely cloomed again.

The bees' entrance, as soon in this chapter is to be shown, must be sometimes larger, sometimes less, sometimes nothing at all. And therefore every beehive must have its gate or summer door, a winter door or wicket, a bar or shutting of the wicket.

The gate or summer door must be made of that size, that the bees in summer, when their number is greatest, may have air enough, with free egress and regress, not hindering one another. The space of four square inches is sufficient for any stall.

This summer door is made thus: first cut away the lowest roll for the space of five inches, and with the briar or thread which bound that part, make fast both ends. Then fill up again the two extreme half-inches of the space, with two door posts.

The door posts are two spleets half an inch broad, and five or six inches long, of which the lowest inch is twice as thick as the other, with a shouldering on the inside. These posts are forced up through the middle of the rolls in their place, to the shouldering, as they serve to size out the summer door to its due space of four square inches; so are they fit to receive the winter door, when it shall be joined to them.

If the hive is with the least, you may set up the posts without cutting the roll.

In a wicker hive the summer door is made more easily.

Sometimes, when a hive is reared, moveable posts are required, which may serve also at other times. A moveable post is an inch square piece of wood, with a

---

36   *Bracks*, breaches, gaps

shouldering above to rest against the hive, and another in the inside of the door to fit the wicker; the form is thus:

The winter door or wicket is made of a piece of wood, an inch and a quarter thick, almost an inch high, and five inches long. At each end of which cut away half an inch all save before, where that half inch in length must be at least a quarter thick, with its full height to fit the door posts. Then in the middle of the underside, cut through the thickness, a hollowness or passage, almost half an inch high, and three inches long, and then there will remain at each end of the hollowness half an inch uncut, besides the two extreme half inches less a quarter thick, and fitted to the posts.

The shape of which wicket you may see in this figure:

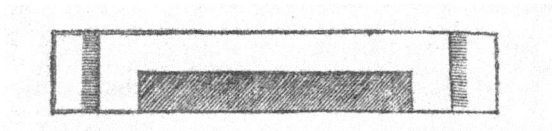

The use of the winter door is to straighten the passage when less room is needed, so that the bees may the better keep out the robbers, that the cold may have the less force, and that the mice may not enter, which in winter are prone to make much spoil.

The bar or shutting is to be made four square of some heavy matter, namely, lead (that neither the rough wind nor crafty titmouse may remove it) in length, depth, and thickness fitting to the wicket, with some little hollowness next to the stool, that may let in the air, and not let out the bees.

For want of lead or other metal, you may with a hammer and grindstone fit a tile-shard, but let that be somewhat broad, that it may lie firmly on the stool.

With this bar you may shut or half shut the wicket, as you shall see cause, to defend the bees in the more dangerous times from frost, snow, titmice, and robbers.

For small stalls, the gate, wicket, and bar may be all of a lesser size.

It is also convenient for each hive to have its bench[37] before it, which may be a plank of the breadth of the stool, and of that length that it may stand leaning from the ground to the forepart of the stool, on which the bees may settle when they come weary or thronging home, and so ascend to the door, and that they may sun and refresh themselves, being chilly and weary. Otherwise you may make a narrow plank or board to serve, fitting the length of it to the breadth of the stools, and then the one edge leaning to the forepart of the stool, let the other be borne up with two forked stakes set fast in the ground, or by some other props.

Bee hives being thus fitted with all necessaries, are afterwards at different times of the year to be differently ordered.

The bee[38] year is most fitly measured by the astronomical months (which begin with the sun's entrance into the several signs of the Zodiac, and are therefore calculated by their names) because as the sun, entering into the twelve signs, and so beginning these twelve months, does notoriously alter his course, making the days longer or shorter, the air warmer or colder, and the earth more fruitful or barren, making also both the equinox and solstice, in which the four quarters of the year, spring, summer, autumn and winter take their beginnings; so the most notable alterations about bees, in things either to be observed in them, or to be done for them, do likewise fall out in the beginnings of these months.

But the four quarters the bees begin one month sooner than the astronomers. For their spring or first quarter begins with Pisces, when the sun begins by his quickening heat to revive the flowers, which all the dead of the winter lay buried in the ground, and the bees having tasted of them begin to breed, and to increase their companies for the fruits of ensuing summer, which from the former summer up to this time have daily decreased. The other spring months are Aries and Taurus.

Their summer likewise contains Gemini, Cancer and Leo, most rich and plentiful in flowers and dews, with which the multiplied bees do now store their cells against the penuries of winter.

Their autumn or harvest has Virgo, Libra, and Scorpio, in which the bee masters and the master bees reap the ripe fruits of many bees' labour.

And their winter consists of the three still months, in which the bees live altogether upon their summer store, and get nothing.

Here note, that although winter and summer do properly betoken two of the four quarters of the year, yet sometimes they are taken, according to the common account, for two half parts: the one containing the warmer season, as from the end of Aries to the end of Libra, the other the colder, as from the end of Libra to the end of Aries.

---

37   *Settle*
38   *Melissaean*

The spring having replenished the hives with plenty of bees, the summer is ready with its plenty of honey to entertain them. During which season the hives must have their largest entrance, lest the thronged multitudes are pestered for want of air, or hinder one another as they go and come earnestly in their work, or are delayed in swarming when they should surpass at pleasure. Neither can the openness of the hives be hurtful to them, seeing now there is no fear of enemies.

At Gemini therefore set the doors wide open, without bar or wicket, and so let them stand all this quarter.

Gemini being past, if the weather is usually cool, when there comes a calm warm day, take off the hackles from those hives that are likely to swarm. But if the weather is extremely hot and dry, then it is good to keep on the hackles to cool the hives.

At mid Cancer double the stalls that lie out. When you would have no more swarms, as namely after the first blowing of blackberries, which is commonly within seven nights after Midsummer, set up those hives that are full with three tile-shards, or other things of similar thickness, and cloom up the space between the hive and the stool. If yet they chance to swarm, as soon as they are hived, put them back to the stock.

Also rear the swarms that being under-hived do lie forth, with bolsters of that thickness that may but let in the bees.

In Leo, or presently after the last swarm, kill the drones of those stalls you mean to take, with a drone-pot cloomed to the door.

And if you see any other so pestered with multitudes, that they are loathe to meddle with them, you shall do well to help them some warm afternoon, and then will they take the work out of your hand, and spend the less time about it.

To the plentiful summer succeeds the wasteful autumn.

At Virgo therefore, or a little before (which is the most dangerous time for bees, because of wasps that then, if not sooner, learn the way into the hives, but chiefly of robbing bees, which then begin to spoil) to the gates of the weaker stalls (whether they are small swarms, or stocks that have cast twice and late) set up the winter doors, and fasten them with good cloom and see that the hives are closed in all places. (Those that have bees lying outside or otherwise are very full, you may let alone and not straighten their entrance until the weather is colder, for such are safe enough.) But first view your swarms, whether they fit their hives; those that have not now wrought down within a handful of the stool, if you mean to keep them (to the end they may lie warm the winter following, and be ready at the doors to keep out robbers) cut off so much of the skirts as will serve the turn (the bigger the hive is, the more you may lessen it) and so set it down, cut a summer door in the skirt, and put to the winter door. Without such help the cold will kill many, and weaken all, by which they become listless[39] in all their doings…

Moreover, because the wasps and robbing bees will soon be stealing, before the true bees are stirring, it is good in the evening, when the bees are all in, to bar up the wickets of those that are weak, that a bee cannot pass, and not to open the same the next day until the weather is warm, and the bees offer to come abroad, though it is not before nine or ten or eleven o'clock, and then you may either open it, or half open it, according to the flight of your bees.

The stalls which you reared in the end of Cancer for fear of swarming or want of room (now that the death of the old bees and of the drones has made room) are to be set down again, lest their swarming be hindered the next year, unless they are swarms that have wrought down to the stool.

Also in this month, about the middle, those hives which you deem to be weak because the bees are gone up from the door, knock with your hand, one after another: they that at the first or second stroke make a great noise both above and beneath, continuing the same for a space, have a store of bees, and are therefore in less danger, but those that make a little short noise, though they are heavy and have honey enough (such as are commonly those of three years old and upward, that have cast twice or oftener that year, and did not by Virgo bear away their drones) yet have they but few bees, and are therefore ill able to resist the violent multitude of robbers, which, when they perceive their weakness, will never leave them, as long as there is a drop of honey in the hive.

If you see them once fighting, either take them without delay, or make their entrance so narrow, that but one bee may pass at once, and before Libra be sure to take them. For though they escape this robbing time through your care and diligence, yet at the spring they will surely yield, or die of themselves, or fly away. Note yet, that whole stalls which are very full, will make but a little noise when you knock them (but different from the other, as being quick, smart and all over the hive) until towards the end of this month, when they are gone up from the door, and their number is somewhat diminished.

In the end of this month is the time to kill and drive bees. Some bees fail after Virgo, and therefore it is good to make trial of them in Libra also, by weighing and knocking the hives; for as they that then make a little noise will die for lack of company, so they that are light will die for lack of meat. And always have an eye to those that the robbers do eagerly run after, which is a sign that they perceive in them some defect or other and therefore will not be answered without their errand.

---

39   *Unlusty*

Such as by these means you find unlikely to live, take or drive: those that you suspect, and yet are willing to keep, mark them, feed them in due time, and prove them again in Pisces and Aries.

At Libra, or before if you see cause, set up the winter doors of the bees, and then diligently in the evenings shut all those in with the bar, that have left watching at the door. For in the cold mornings, while the true bees keep in, because it is not a fit time for them to gather in, the thieves, both wasps and bees will be abroad, seeking where they may break in and steal. But still let the weaker have their wickets half shut.

This shutting and opening of the wickets must be continued throughout Scorpio also, unless abundance of cold rain sooner chastens than the wasps. But for the poor stalls, it is best to keep them half shut all the day long, as in Virgo and Libra.

At Scorpio dress your hives for winter. First lift up the stalls (except those that are full of bees, which will not need your help) and sweep the stools clean, then setting them down again warily, that you hurt no bees, cloom them close, and mend all breaks and faults about them, and where the hackles are worn, set new in their steads, that they may keep the hives dry and warm. And now remember also to shut the wickets of them all.

After autumn, the sun drawing near the winter tropic, with a short and low course above our horizon, there follow three still months, Sagittarius, Capricorn, and Aquarius, in which as the plants lie still in the earth waiting the sun's return to revive them, so the bees lie still in their hives, passing this fruitless time in sleep and slumber. Yet so, that if there happens a mild and warm hour, they soon perceiving it, awaken out of their slumber and hasten out of doors with all alacrity, that they may take the fresh air, recreate themselves, drink, exercise their wings, carry out their dead and other unpleasantness, and lighten their little bellies, which are often so stuffed, when the weather allows them not to go abroad, that they can hold no more, so loath are they to defile their nests. And having thus refreshed themselves, at their return, they take their repast, and then return again to their rest. But many such days, especially in times of scarcity, are dangerous, causing them to spend much of their store, which in still frosts they would spare.

The first foul and cold weather in Capricorn shut the wickets close, to save the bees from the titmouse, and from the cold, as well within the hive as without. For as the frost and snow and cold winds, yes, and the ordinary disposition of the air chills many of them, whom the flattering sunshine entices abroad, so the great frosts[40], striking through the door, freeze the lowest bees in the hive to death,

---

40   Seventeenth century winters were harsh; the first recorded frost fair on the River Thames took place
     during the winter of 1607/08, the year before the first edition of the *Feminine Monarchie* was published.

so that by little and little many stalls in some winters have been thereby wholly destroyed, which, by keeping them warm, might have been preserved. But when you shut them in, be sure the hives are always surely closed, for the bees when they awake will strive by all means to come forth, though they never find the way in again. Yet when there happens any pleasant day (namely when the sun shines, the wind is still, or blows mildly out of the south or west, and the earth is without frost and snow) it is expedient to give them leave to play and to refresh themselves: once in a fortnight or three weeks is to be wished, especially after Capricorn is passed. But if you or the weather shut them in much longer, they will be so faint and feeble through their long restraint, that without very pleasant weather at their coming abroad, a number of them will be chilled while they rest themselves but a little in the open air. And therefore as often as, for this purpose, the door is a little opened, alter it not, until the weather alters, and when Aquarius is half spent, if, for fear of a piercing night-frost, you bar them up in the evening, let them go again in the morning, unless either snow or boisterous winds forbid you.

The still months of winter being passed, the new year enters with Pisces, the first month of the spring, when the plants begin to sprout, and the bees to breed again.

Now therefore, if not sooner, the weather being fair, half open the wickets of the better sort, and so let them stand day and night. For the night-cold, being now shorter and weaker, is not dangerous to such, and the day-cold does them more good than hurt, causing them to lie still and spare their store, until it is a fit time to go abroad. But for the weaker swarms (which are more subject to cold, and robbing that now begins afresh) shut them close in the evenings, and open them not in the mornings until it is warm, and then give them but room for a bee or two to pass, especially those that stand most warm in the sunshine, which makes the robbers able to endure the siege, whom otherwise the air's chilliness would quickly discourage.

And now (the bees beginning to breed) is the time to dress and fill their drinking troughs, which all the winter lay neglected.

At this time, in a morning before the bees come much abroad, lift up your hives, and quickly sweeping the dead bees and other offensive material away, and scraping clean the stools, set them down again, and cloom them close as before. For although the bees in time would rid them clean themselves, yet shall it be good for them to have it done at once, that they are neither hindered, nor annoyed by that, and now and then the carrying out of a dead bee at this time of the year does cost a quick bee her life, for being drawn with the weight of the corpse to the cold ground, while she stands panting a little, she is chilled, and so is not able to rise any more.

This cleansing of the stools, after a calm Aquarius, when the bees have been much abroad, is not so necessary, and especially for the better stalls.

Those that by their lightness you perceive to lack honey, you may now save by feeding or driving them into others that have store.

Aries is almost as dangerous a month, for robbing, as Virgo, and therefore you must have a care in the evenings to shut the wickets, and in the mornings not, before it is warm, to half-open them again and where the dry winds and hot sun have shrunk the cloom, be careful to fill up the chinks again.

The poor stalls this month would be half-shut all the day, as in Virgo and Libra.

At Taurus, and sooner, if sooner you see cause, removing the bars from the better stalls, set the wickets open: and for the weaker sort, let them all this month be shut in the evenings, and in the mornings, as soon as it is warm, be but half-opened.

At Gemini take away the wickets from the better, and the bars from the weaker stalls, and when this month is half passed, make them all alike, leaving the doors as they were in Gemini before.

Chapter 4
Of the Breeding of Bees,
and of the Drone

The drone, which is a bulky hive bee without a sting, has been always reputed a greedy scoundrel[41] (and therefore he that is quick at meat and slow at work is fitted with this title) for however he braves it with his round velvet cap, his side gown, his full paunch, and his loud voice, yet he is but an idle companion, living by the sweat of others' brows. For he works not at all, either at home or abroad, and yet spends as much as two labourers; you shall never find his mouth without a good drop of the purest nectar. In the heat of the day he flies abroad, aloft, and about, and that with no small noise, as though he would do some great act, but it is only for his pleasure, and to get him a stomach, and then returns he presently to his cheer…But for all this there is such necessary use of him, that he may not be spared, for without him the bee cannot be.

The general opinion in reckoning the drone is that he is made of a honey bee that has lost her sting, which is even as likely, as that a dwarf having his guts pulled out, should become a giant. Others seeing the foolishness of this opinion, have thought and taught that the drone is a different species, and that as bees breed bees, so drones breed drones, which thought (if the author had observed, that at the time of their breeding and many months before, there is not a drone left alive to breed them) he would have liked as well as the former. These opinions then, being one as likely as another, let them go together. The truth is, they are of the same species as the honey bee, but of a different sex.

[42]For though it is true he is not seen to conceive with the honey bee, either abroad, as other insects do, or within the hive (where yet you may by no means behold what they do) yet without doubt he is the male bee, by whose natural heat and masculine virtue the honey bee, which breeds both honey bees and drones, secretly conceives.

The reasons that move me thus to think, are these. First, because although they are great wasters of the bees' store, yet until they begin to leave breeding, and have conceived for the next year (which some do about Leo, most before

---

41   Lozell
42   Butler's methodology is made very clear in this section. He first lists some of the common folk lore about drones, and then he comes to his conclusion based on observational evidence and deduction. It was, however, Jan Swammerdam in Holland (1637 – 1680) who gave the first confirmation from microscopy that the queen bee has ovaries and is the sole mother of the colony, even though he was puzzled by the exact mode of reproduction: "I do not believe the male bees actually copulate with the females." Biblia natura, 1737

Virgo) they tolerate them; afterwards they begin to beat them away. Which if some do not, before Scorpio they die naturally, and from that time onwards all the winter, until the bees breed anew again, in reality there is not a drone to be had, in the nature of things. When they are quite gone, then do the bees lay no more seeds that year, but only hatch and breed up those that are already in the cells.

Secondly, the earlier and therefore the more drones there are, the more and greater are the swarms...so where the drones are few and late, there is small increase, and therefore if you kill the drones of a hive before the bees have done swarming and breeding (as some foolishly have done before mid-summer, to save their honey from these lazy gluttons[43]) neither will the swarms come forth that were formerly bred, nor the stock from this time on breed anymore. After which time bringing in bee-pollen[44] as much as before, and having no young ones to spend part, they lay it up carelessly in their cells, where it corrupts and turns to a stinking blockage, which will cause them so much to dislike their hive, that the next Virgo they will easily yield to the robbers. And if by your industry they are then preserved, in Pisces, when breeding time is, finding their wombs barren, and therefore loathing even themselves and all, they yield their goods to those that will take it, and after a while, when the strange bees and they smell all alike, by conversing together in the same hive, and sucking the same honey, away they go with them to their drones. But every fair day they will return to fetch that which they left behind them; you may see them fly so thickly to and from that hive, as if it were full of bees, but when night is come, they are all gone.

Thirdly, because every living thing does breed male or female of its kind, and experience does teach us that the bees do yearly breed, as well drones as honey bees, seeing the honey bees are females, it follows necessarily that the drones are the males of the same kind. And therefore in the learned languages the drone has his masculine appellation, as the honey bee her feminine.

Fourthly, we see the same in similar insects, the wasp and the bumblebee, for the demonstration of which I will briefly show you the breeding of them both.

The wasps' nest is begun by one great wasp, which you may therefore call the mother wasp, which in Cancer (or when hot and dry, it springs somewhat earlier) in some hole, usually made in the ground by a mole, mouse, or other means, works a comb of the decayed wood or other timber, in the form of a round tent hanging by the top to the outer part of the hole. This comb contains about six cells, of the size and shape of the bees' cells, in which she breeds so many

---

43 Lurchers
44 Ambrosia

young ones, which, when they are fledged, breed as well as their mother[45], and so enlarge the comb to some eight inches over. Then, making more room beneath by moving and carrying out the earth, they hang another comb under the first, by little pins, and so another, and fifty others, increasing still in the same place until summer is done. For they go not forth in swarms as bees do…When their breeding draws towards an end, namely in Virgo and after (besides the small or ordinary wasps, which lie in all the upper combs) in the last or lowest comb, made for that purpose with larger cells fit for larger bodies, they breed also two other sorts, drones or male wasps (which are somewhat bigger and longer than the small wasps, and without stings as the drone bees) and mother wasps, which are like the small ones in all respects, save that they are twice as big. These when they are fledged having been conceived like the bees, by the drones in Libra, and sometimes sooner, fly abroad (as their drones also do) gathering for themselves, and searching and prying into every corner as they go, for their winter lodging; and after a while, when the air waxes cold, leaving both drones and small wasps to the mercy of winter (which with its first cold-wet weather chills and kills them as they fly abroad) do forthwith take themselves to some warm place, like the thatch of a house, a mortice in a post, an auger hole, or the like; but especially into hollow trees (which is the reason why in grounds adjoining woods their nests will be most rife) where they abide until the next spring without any meat, as it were in a deep sleep, out of the which nevertheless a little warmth of the fire, or of your hand will awake them at any time. At the blossoming of willow trees, if the weather is warm, they fly abroad for food, and in Cancer or Gemini, as I have said, they begin to nest and breed. He that kills one of them, kills a whole nest of wasps.

That the drone wasps are the males, is the opinion of some in the days of Aristotle, as he himself writes…[46]

The bumblebee[47] likewise begins her nest single, being more like the bee than the wasp is, in that she makes honey, and more unlike in the shape of her comb, for she has only a few round cells of the size of grapes, lying flat on the ground one upon another without any order, which are made after this manner. First either upon the ground in the grass, or in some shallow hole in the ground, she prepares a little stuff which is soft like wax, but brown and more brittle, of the size of her head, and in which she lays about six or seven seeds together, circling them round with the same stuff, which increases by little and little as the seeds do, and when they begin to live, it grows into as many several cells, as there are grubs, each one having one to himself. When they have grown, the cells, which before were brown and brittle, now wax white and tough, that you can scarcely

45 Dam
46 Aristotle, *The History of Animals*
47 Doore

tear them. And when the bumblebees are ripe, they gnaw their way out at the top. Upon these they make more in similar manner, and the empty open cells they fill with honey, with which they feed both themselves and their young, when the weather allows them not to fly abroad. All this nest is covered with a little moss like a bird's nest. Until Leo they breed females only as the wasps do, and then last of all, for propagation of their kind, they breed their drones, being likewise, as the drones of bees and wasps, without stings. And these, to put the matter out of doubt, within a month after they are ripe, openly procreate with their females, as the beetles[48] do, but their mates they choose in the nest, and are carried away by them. After which time the females breed no more until the next summer, though you may see them gathering, and flying about somewhat longer than the wasps. In Sagittarius they take themselves to their winter rest, where they lie single as the mother wasp does in a sleep. But the drone bumblebees, as the drone wasps, are destroyed by the weather: not one afterwards is to be seen until next Leo, when the females breed anew again. But one thing in the bumblebees and wasps is more strange, than in the bees. For whereas the bees as soon as they have bred their first brood of females, do soon breed drones, (which both, which when they are ripe, multiply together) the young bumblebees and wasps in the beginning of summer, do not immediately take the drones (for then there are none) but receive from their mothers, together with their nature and being, that masculine seed, whereby when they are ripe they breed all the summer following, until in the end they likewise conceive by their late bred drones for the next year, both for themselves and the young that shall come of them.

By this time you will say with me, that the drone is the male bee, for which reason if some curious surgeon[49] would make an anatomy, he should early discern…two lawful witnesses of his masculine sex.

This truth began to appear many years ago, even in Aristotle's time…which opinion he recites in another place…Where though he does not approve it, yet has he no other reason against it but this…Nature has armed no female for fight and force against the male, but the bees have the power and weapons to chastise the drones and therefore the drones cannot be their males[50].

The weakness of such reasoning I marvel he did not see, seeing in all the kinds of hawks the female does command the male, as being both stronger and better armed. Unto which may be added the example of the Amazons reigning in his time, who by force of arms subdued many kingdoms of men, and held them in subjection, compared to which it is a marvel but there were then some Viragoes[51]

---

48   Cockchaffers
49   Chirurgion
50   i.e. according to Aristotle, since the female bees have coercive stings, and the drones do not, the drones cannot be males since this is against nature.
51   Female warriors

in Greece, as well as there are now in other countries. Which, if nothing else, the experience of his masters[52] might have taught him.

But you must understand that the Philosopher speaks thus, not dogmatically but disputatiously, only by way of reasoning, for in the end of the same chapter he yields himself to have no certain knowledge of this…

To return therefore to our purpose, the honey bees having been, as those other insects, conceived by the drones, the strongest ones about Pisces, when they first gather upon flowers, others in Aries, and the weaker ones later, they begin their breeding, which is continued all the summer, even to the end of Virgo. But the chief time is in Aries, Taurus and Gemini, which months yield bee-bread[53], the young bees' food, in greatest plenty, variety and virtue.

The bees will be sure to serve themselves first, their first generation being always females, which they breed after this manner.

Close under the honey (which is at that time altogether in the upper parts of the combs) in the middle of the bottoms of the empty cells, as the wasps do on the one side, they lay their seeds, about the size of those which the butterfly leaves upon the cabbage leaves, but of different colour, the bees being white like wasp seeds, and the butterflies yellow. And so they descend by degrees toward the lower part of the combs, filling one cell after another. Although when the chief breeding is past, they do not precisely observe this order, but lay up their honey promiscuously among the young bees, where they find the cells empty. The bee seed at the first sticks upon one end, until it becomes a live worm or grub; as soon as it lives it is loose, and lies in the bottom of the cell round like a ring, one end touching the other, until the bottom can no longer contain it; after that, it lies along in the cell until it is grown to the full size of a bee and then does the worm die, and becomes void of all motion and sense, and so is shut up in the cell, the bees covering the top shut with wax.

The grub being now dead, presently begins the alteration from a worm to a bee, which is two-fold, in shape and in colour. The first alteration in shape is the division in the middle, then the other division between the head and shoulders, when it is called an insect; after that, the growth of the head, legs, wings and other parts into their shape and fashion. The first that alters in colour from white to brown is the upper part, and of the upper part the head, and of the head the eyes.

The uniform shape and white colour of the worm, being thus altered into the proportionate shape and brownish colour of the bee, she begins to move again, and to live out her second life, and then breaking the cover with which she was enclosed in the cell, she comes forth a flying bird…And all this within the space

---

52   Butler's footnote suggests these were Plato and Socrates
53   Bee-bread, bee-pollen

of a month. In swarming time, when the hives have more heat, partly from the air, and partly from the multitude of bees, when also the young bees never lack their fill of nectar, pollen and fair water continually brought in fresh and fresh unto them, I have known this effected in three weeks, although Pliny[54] speaks of more than twice so long a time.

But the lady bees are bred in the several palaces of the Queen after a peculiar and more excellent manner. For the golden matter of which they are made, is not turned into a worm at all, but immediately receives the shape of a bee…

When the old bees have ended their first broods of females, then last of all after the same manner in wider cells made for that purpose they breed the male bees or drones, as was long since observed…And therefore some stalls do not produce drones before Cancer, not many before Gemini, nor any before Taurus, although you may see the nymphs of good stalls abroad in Aries, of others in Taurus, and of all in Gemini. By chance some few young drones may be bred in good time with the females, but they, coming out of season, are not allowed to live.

These young drones, when they are fledged, do not only serve for reproduction (as has been shown) but also help the females greatly, by reason of their great heat, in hatching their broods…And for these reasons they are always in breeding time mingled with them throughout the hive. Although afterwards (when they have been much beaten, and can go nowhere single, but one or other will be on their own) they gather altogether in a cluster, for their safety in one side of the hive, so that it is true at some time of which the Philosopher spoke indefinitely, 'They take the place in the deepest part of the channel [55]. And yet their hanging together will not serve their turn, for the bees, when they are disposed, will quickly make them part, and depart. When there is no use for them, there will be no room for them.

For the drones are but vassals to the honey bees, which as they do excel them in virtue and goodness, so do they also in power and authority, ruling and over ruling them at their pleasure…For albeit generally among all creatures the males, as more worthy, do master the females, yet in these, the females have the pre-eminence, and by the grammarians' leave, the feminine gender is more worthy than the masculine…But let no nimble tongued reasoner gather a false conclusion from these true premises, that they, by the example of these, may arrogate to themselves the like superiority…and he that made these to command their males, commanded them to be commanded. But if they would so pretend, have it so, let them first imitate their singular virtues, their continual industry in gathering, their diligent watchfulness in keeping, their temperance, chastity,

---

54   Pliny the Elder, CE 23 - CE 79 *Natural History*
55   Aristotle, *The History of Animals*

cleanliness, and discreet management. And then, if they meet with such dull idlers[56] as these drones are, they may with less blame borrow a point of the law, and enjoy their longing. Yet when they have it, let them use poor skimmington[57] as gently as they may, especially in public, to hide his shame.

And this they may note by the way, that even though the females in this kind have the sovereignty, yet have the males the louder voice, as it is in other living things, including doves, owls, thrushes etc, the males being known by their sounding and shrill notes from the silent females. Yes, the wives themselves will not allow that hen to live which presumes to crow as the cock does, nature teaching, that silence and soft noise becomes that sex.

The bees breeding or laying of seeds begins to cease, in some by Leo, in some not before Virgo. After which time these Amazonian Dames, having conceived for the next year, begin to wax weary of their mates, and to like their room better than their company. At first not quite forgetting their old familiarity, they gently give them Tom Drum's [58] entertainment; they that will not take that for a warning, but presume to force in again among them, are more shrewdly handled. You may sometimes see a handful or two before a hive, inside which they had killed, but the greatest part flies away, and dies abroad.

But because in the same hive they do not leave breeding all at once, therefore neither do they kill their drones all at once, but at first taking away only the superfluous, they allow as many as they need, to remain longer – some often a whole month longer.

The forward stocks, that have cast their last swarm in Gemini or soon after, begin at Leo: yes, of those in the beginning of Gemini some somewhat sooner; the backward, that cast not their last swarm much before Leo, may stay until the end of the same month, but usually about Virgo, or a week after, they make a clean riddance of them.

Those stocks that being full have not swarmed at all, because they are rich and fear no want, used to allow them so long and sometimes longer, even to the end of this month. Those that have over-swarmed themselves, finding their paucity and weakness, become desperate and careless of their estate, and therefore sometimes keep their drones until towards the end of Virgo; sometimes they kill them not at all, but let them alone, until they die by nature, which is not long after. For few of them can live till Libra, and the youngest not to the end of that month. Take heed to such stalls, for they are likely to die.

---

56   Lubbers
57   A *skimmington* was a folk custom in which villagers noisily paraded an effigy of the person they wished to humiliate and of whom they disapproved.
58   Popular entertainment

Some are so provident, that, to prevent this trouble and save their honey, they draw the poor young drones out of their cells before they are ripe, or come to their second life. Such you may safely trust.

Those that the soonest rid themselves of their drones, are likely to be the most forward the next year.

Sometimes the drones are beaten away in the spring. For when forward stalls (which in their heat are bold to fly abroad when others dare not stir[59]) have lost many of their nymphs in a tempestuous and stormy spring, they will therefore destroy their drones also. But having formerly conceived by them, they then begin the world anew, as after another winter, and first breeding nymphs, in the end they breed young drones again. Which if they can compass before swarming time be past, they will swarm that year, otherwise they will be fat and full, and good, either to keep or kill.

Because the stocks that have cast often, bear with their drones so long, although there are twice so many as bees as needed for the bees that are left, therefore (to save the honey which those gluttons would devour) it is not amiss to prevent the bees, and soon after the last swarm to diminish their number, with a drone pot cloomed to the door, especially of those you mean to take, or otherwise you will be much oppressed with a superfluous multitude.

---

59   Wagge

Chapter 5
Of the Swarming of Bees,
and the hiving of them

The stocks having bred and filled their hives send out swarms. A swarm consists of all such parts as the flock does: namely, a queen bee, honey bees old as well as young, and drone bees.

If any man desires to see the queen, he now has opportunity, when she goes forth with her swarm and many dead ones he may find before the stools[60], when the flocks have cast their last swarms and also when many meet in one swarm. But then, being dead and shrunk together by the force of the poison, they lose much of their stature and comeliness.

Men think that the swarm consists only of young bees, and that the old bees only tarry behind, but indeed (though it may seem strange) the swarm is no younger than the stock, for there are in both of both sorts. The young bees remain in the flock with the old for their defence, and for the greatest labours, and the old ones go with the young in the swarm for their aid and guidance in their work.

The drones they take with them for propagation of their kind and therefore those swarms that have many drones will surely prosper, and if they are early[61] will swarm again, unless they are over hived, whereas those that have few or none, will increase little or nothing all the summer after.

A warm, calm, and showering spring causes many and early swarms though sudden swarms do hinder them... Dry weather makes plenty of honey and the most swarms. But note that the chief time for breeding swarms is the spring and for honey gathering, the summer, so that when a dry summer follows a wet spring, the bee-folds are rich. If the summer is also wet, the increase of bees will be greater, but, because of the scarcity of honey, this increase will produce a decrease; the more swarms you have at the end of this summer, the fewer stalls shall you have at the beginning of the next. Except for some fair early swarms, and some good stocks, which cast at an early period or not at all, they all die because of hunger, when they have spent their own pittance, and spoiled their own fellows...To prevent the loss and spoil that would come of that, take the light stocks, together with the small and late swarms, feed the middling sort, and be sure they are not over-hived.

Likewise in warm and calm weather, the swarms delight to arise, but especially in the warmth of the sun after a shower or gloomy cloud has sent them home

---

60   Stools – the stands on which the hives are placed
61   Rathe

together; in extremely hot and dry weather not so, in so much that stalls being full and ready to swarm with the first, are sometimes so kept back with cold dry winds in Gemini, and with extreme heat and drought in Cancer, that they have not swarmed at all that year.

The swarms' custom is to come forth between the hours of nine and three, and sometimes an hour sooner or later, but chiefly between eleven and one. They choose rather the forenoon, if the weather pleases them, otherwise they will stay for a fair hour in the afternoon. This time of the day therefore, in the swarming months, your bees must continually be attended.

The swarming months are two, Gemini and Cancer, one month before the longest day, and another after.

Those that come before the solstice, in the ascending of the sun, are early swarms. Those that come after, in the sun's descending, are late swarms. But there are a few that come in the first fortnight, and they very good; a few also in the last fortnight, namely after St Peters-tide, and they all are bad, unless the backwardness of the year, when it happens, mends them.

Those that swarm before the blowing of knapweed, come in very good time; before the blowing of blackberries, they may hive and do well, but blackberry swarms, especially castlings, are seldom to be kept, as being more likely to die than to live, and if they hive, they seldom swarm the next year. And moreover they weaken the stocks from whence they came, which otherwise the next year would swarm in good time, and then one such swarm is worth three of those late ones. In which case, put them back again into the stock, which you may easily do, so soon as they are hived, by knocking them down upon a table close to the door; their fellows that are behind will soon be in with them. And if they rise again, serve them so until they cease. But if you spy them rising before the queen has come forth, shut them in a while, and that will stop them.

A good stock naturally and usually casts twice: a prime swarm, and an after-swarm, especially if the prime swarm is so early, that the casting comes before the bramble buds are open and early prime swarms over hived, in a plentiful year may swarm once or twice, although some full stalls do not cast once, some but once, and some having many princes (especially when the prime swarm is broken) cast three or four times. For sometimes it happens that, in the swarming, a rising black cloud stops part of those which have already come forth, and they lie about the hives door, sometimes when they are all up, either fearing a cloud, or disliking the alighting place or being troubled in the hiving, part of them return.

One prime swarm is worth two or three after swarms, unless it is broken, and then if the residue comes forth in one entire swarm, that after swarm may be the better of the two, but if it is divided into two or three, then will they all be indifferent, so much so that unless they are timely, or united, they can hardly live until the next summer.

The choice of the time when the first colonies, or prime swarms go forth, the rulers refer to the commons[62], who by reason of their continual travel and business both without and within, do best know when all things are ready and fit for them. First within they will be sure that they have a prince ready to go with them, for without a governor they will not be[63]. Then that their hive is full, so that it may be divided at the least into two or three sufficient companies, one to remain with Marpesia[64] the old queen, another to go forth with Antiope the prince[65] and a third perhaps, which together with the unripe brood in the cells, may make another swarm to serve Orithyia[66]. Outside likewise they will see, first that the flowers are sufficient to furnish them with a store of wax and honey, then that the weather pleases them, as being warm and calm, and moist, unless, being continually unseasonable, they have no choice.

When the hives begin to be full, they will produce drones, or yield fledgling drones, which is a sign that the first brood of nymphs have been a good while flying abroad, and are now able to endure both weather and labour. Other signs of the hive's fullness and readiness to swarm are at the hive-doors. First, the bees hovering in cold evenings and mornings. Secondly, the moistness or sweating upon the stool. Thirdly, their hasty running up and down. Fourthly, their first lying forth in foggy and sultry mornings and evenings, and going in again when the air is clear.

When they do swarm, sometimes they first gather together outside at the door, not only upon the hive, but upon the stool also, where when you see them begin to hang one upon another in swarming time, and not before, and to grow into a cluster that covers the stool in any place (especially if there are drones among them) then be sure they will presently rise, if the weather holds. The first that come forth will increase that cluster to some fourth part of the swarm, and then they will begin to fly away, first out of the hive, and after from the cluster. But commonly a few of them first fly forth and play to and fro at the hive door, so to draw out more company to them, and when by this means they have got out so many, you may see them begin to dance about the hive, and then do they hastily issue forth and swarm.

---

62 Commons – the common people, hence the inhabitants of the hive
63 The political parallel is made between a human society, well governed and ordered, and the bee community which offers, in Butler's view, an example of an harmonious society where each member works towards the common good
64 In ancient Greek and Roman legendary history, Marpesia was Queen of the Amazons
65 Antiope was an Amazon, daughter of Ares in Greek mythology
66 Antiope's sister

But here you must note, that as to filling the door, or lying out a little in foggy or sultry mornings and evenings (which is because then they are most offended by the heat inside, and can best endure the air abroad) and otherwise to go in again, it is a sign that the hive is full, and therefore ready to swarm. So to lie forth continually (as in extremely hot and dry summers they are wont to do) under the stock or behind the hive (especially after Cancer has arrived) is a sign and cause of not swarming. For the bees, knowing by nature, that the greatest companies prosper the most, until they find themselves so pestered with the heat and throng of the multitudes, that the hive can scarcely hold any more, they will have no mind to swarm, and when they have once taken to lie forth, the hive will always seem empty, as though they wanted company.

One cause of their lying forth is stormy and windy weather, not allowing them to swarm when they are ready, for when their number has grown so great through their continual breeding, that the hive cannot hold them, seeing they may not swarm, they must for want of air and room within, lie outside, which when they have once caught, they will hardly leave, and the longer they lie out, the more reluctant they are to swarm.

Another cause of their lying forth, is continually hot, dry weather, especially after the solstice, which causes plenty of honey both in plants and dews, and their minds are so set upon their chief delight, that they have no leisure to swarm (although they might most safely come abroad in such weather, which would not allow the weakest nymph to fall.)

And when by continuance of such honey-weather they are once sufficiently provided, they will then be loath to leave the sweet fruits of their labours, and to change their full store-houses for that which makes giddy housewives. But if they have once begun a comb outside which they lie, the matter is beyond a doubt. Whereas contrarily in wet and scanty summers, no weather will stop them from swarming as soon as they are ready, although by that means (unless they are early, or the weather suddenly mends) most as well of the stocks as swarms are likely to die for hunger, and therefore, as near as you can, so order the matter, that your swarms may come on time[67]. Early swarms and their stocks, that have the summer before them, always prove good. But for those stocks, which not swarming in Gemini happen afterwards to lie forth, the following may be a remedy:

First keep the hive as cool as may be, by watering and shadowing both it, and the place where it stands, and then enlarging the door to give them air (always provided that there is no back door in the shady parts of the hive) move the cluster gently with your brush, and drive them in.

---

67 The management and encouragement of swarms to increase stock is a key component of Butler's advice, in comparison with more recent beekeeping methods which seek to minimize the likelihood of swarms cf. *Swarming and Its Control and Prevention* L. E. Snelgrove, 1934

If still they lie forth and do not swarm (even though they have had suitable weather for two or three days) then the next calm and warm day, between eleven and one of the clock, or within an hour sooner or later (when the sun shines and you see no clouds coming to hide it) put in the better part, at the least, of those that lie out, with your brush, and the rest gently sweep away from the stool, not allowing any to clutter again. These rising in the calm heat of the sun, and flying about before the hive, will make such a noise, as if they were swarming, which their fellows hearing, will happily come forth to them, and so begin to swarm.

If this does not suffice, but that returning to the hive, they lie forth again, then raise the hive high enough to let them in, and cloombe up the skirts, except for the door.

But if not withstanding all this they do not swarm, then assure yourself they have no prince[68] bred to go forth with them, or else they are fat and full of honey, which they are resolved not to leave.

And then if it is before mid-Cancer, and the honey weather hold, your best way is to double the stall, by turning the skirt of the hive upwards, and setting an empty[69] prepared hive fast upon it, into which they will ascend, and work and breed there as well as in the old. In the end of Virgo drive them all into the new hive (which then, if the weather has held good, will be full of wax and honey) and take the old for your labour. But if mid-Cancer and the honey-dews are past (because they need time and means to store the unoccupied hive, let them stand – such a stall will be very good to be taken, or if young, to be kept. But first replenish some over-swarming hive with its excess or lying out (especially if you mean to take it), in this way:

When all hope of their swarming is past, in some evening (while it is yet light) holding a hive under those that lie out, cut them off from the stool with a tight thread, and carrying them to an over-swarming hive that you would mend, knock them down on a table cloth before its hive, into which, because they come without a prince, they are quietly admitted and quickly united under one common commander.

The manner of doubling a stall is this: having first measured the hive about in the largest place, provide an empty spleeted[70] hive of the same size and compass. Make ready also two square sticks, thirteen or fourteen inches long and an inch thick at one end, and half an inch at the other. These two sticks lay parallel over the hives five or six inches apart, and each of them a similar distance from the middle of the hive, with both the thick ends one way to size out the door for this doubled stall, and so tie them with needle and thread to the skirts, fast in their

---

68   See footnote 76
69   Leere
70   Supported by spleets, sticks, to strengthen the skep walls.

places. These sticks serve to keep the hive from slipping, and to save the bees, that otherwise might be pressed to death between the two skirts. Then in a fair night, as soon as it is dark, raise the full hive with three bolsters, two on the west side, and one on the east, some four or five inches high (or with a double rest) to let the bees in, and cover both it and the stool with a large mantle. Then make a brake behind the stool of four stakes[71], two feet and four feet long, pitched fast in distance equal, and fit to contain the whole hive, which you may be sure of by fitting it to the empty hive, being of the same compass. One of the short stakes set close to the middle of the back of the stool, the other northward opposite to it; one of the long ones on the west part, and the other on the east. Then right in the middle, between the stakes, dig a hole in the ground half a foot deep, and of such compass, that being half filled with a wisp of straw, it may fitly receive the top of the hive, and so the hive may stand upright and fast in the break. Then pare away the inner edges of the tops of the short stakes, that the hive in the setting down may not stay against them, and taking up the west stake, lay it by you.

These things thus prepared, yourself standing on the west side of the brake, and your assistant[72] on the same side of the stool at your right hand (both in your complete harness) let the assistant take hold of the hive and yielding the top toward his breast, raise the far-side of the skirt from the east bolster[73]. When you see it fitting, embracing the hive as near the skirt as you may, lift it up sheer from the other two bolsters, and set it down warily in the middle of the brake, with the top in the hole as upright as you can, and the door to the rear part of the stool, that it may stand southward as it did before. And immediately let your assistant, being ready, place the empty prepared hive evenly on it, with the thick ends of the sticks southwards, and then put the long stake into its place again yourself. Then cloom the hives together with rolls, laid flat, so that none of the cloom falls in among the bees, leaving open the space between the sticks' ends for the new door of this double hive. Lastly put on the hackle, and gird both it and the long stakes to the empty hive, about the middle with a belt, and about the top with a band of willow[74]. And so let them stand until after the end of the hottest days of summer[75], when bees are taken. But in no case let the doing of this be deferred beyond the time prescribed, in case you have little or nothing for your labour.

At the harvest season, on a fair calm morning before any bees are abroad, shut up tightly all the stalls in your garden, and those that stand next to them cover with sheets and blankets, lest some of the younger sort mistake, and tarry at their

---

71   Cf. *The Earliest Record of Beekeeping in Northern England* Robert J Hawker, 2015 p 17
72   Presumably a family member or a boy employed to help with the hives. Cf Francois Huber who relied on his wife and a servant to assist him. Huber, *New Observations Upon Bees*, 1792
73   Support made from a stick
74   Withe
75   Dog-daies

doors until they are chilled. And when the sun is an hour high, and the air waxes warm, having first parted the new combs and the old with a long knife, take off the upper hive or receiver, and set it on the stool in the old place. But be sure, if you see that the upper hive is very fat, or if you fear the queen is hurt, or not in the receiver, your best and safest way is to take them both, for if they are over fat or need a ruler, undoubtedly they will not prosper.

The signs of after swarms are more certain. For since the rising of the prime swarm is appointed by the vulgar, whose chief rule is the fullness of the hive, the hive being now well emptied, for other swarms there needs some other direction, which the rulers themselves do give by their voices, without which that stock will swarm no more that year. And yet the choice of the hour, and of the day among four or five is permitted to them, as best knowing the disposition of the weather.

When the prime swarm is gone (if the stock casts any more) the eighth or ninth evening after, sometimes the tenth or eleventh, the next prince[76], when she perceives a competent number to be fledged and ready, begins to tune in her treble voice a mournful and begging note, as if she did pray to her queen-mother to let them go. To which voice, if the queen vouchsafes to reply, tuning her bass to the young princes' treble (as commonly she does, though sometimes scarcely entreated in a day or two) then does she consent. And therefore, unless foul weather stops them until it is too late, you may assuredly look for a swarm. Which seldom arises the next day, although the weather is very pleasant, or the next day, unless, after the third night's warning, they will accept indifferent weather, during which the prime swarms will not come abroad. And as the queen's voice grants the wish, so her silence is a flat denial: the proverb here has no place, *Qui tacet consentire videtur*[77]. For without this consent, there is no consent.

This song[78] being contained within the compass of an octave from C to c-sol-fa, the prince composes her part within the four upper clefs G, A, B and C usually in triple mood, beginning with an odd minim in G-sol-re-ut and tuning the rest of her notes, of which the first is a semibreve, in A-la-mi-re. Sometimes she takes a higher key, sounding the odd minim in A-la-mi-re, and the rest in B-fa-b-mi. Sometimes, especially towards their coming forth, she rises yet higher to C-sol-fa, holding the time of three or four semibreves, more or less. Now and then she begins in double time some two or three semibreves, but always ends with minims of the triple mood.

---

76   A female prince, referred to by Butler as 'she'
77   'he who is silent is taken to agree'.
78   'The clergyman's excursion into bees' music is, obviously, an oddity and it appears to this writer might well have been written with tongue in cheek.' *Charles Butler – Musician, Grammarian, Apiarist* by *James Pruett* in *The Musical Quarterly*, Vol 49, Pt 4 (1963)

The queen's part, contained in the four lower clefs, consists of minims altogether in triple mood, commonly in sa-sa-ut, sometimes in C sol-fa-ut, sometimes in the other two clefs between them, continuing her tune in the time of nine or ten semibreves, more or less.

Sometimes a third princess, imitating the queen's voice in time, though differing perhaps in tune, joins with them, the more with their full noise to incite the swarm to go, that her turn may come the sooner. And sometimes a fourth also interposes her minims to fill up the choir. But none dare interpose the voice of the chief prince, for that were to be treason to her person (and yet sometimes one of them, in the hope of parting the swarm, will steal out with her) which, if the swarm is not parted, or being parted is put together, costs her life, as well as the lives of some of her followers. Notwithstanding each of these, when her elder sister is gone, and her turn next, changes her note, begging in Orithyas' tune leave to be gone, too, which as sometimes the queen grants unasked, beginning first herself, so sometimes by her silence she denies, though mournfully entreated, and then the swarm tarries, and the poor lady must die.

With the various and harmonious notes, answering one another, and some pause between, they go solemnly about the hive, so to give warning to all the company. This they continue daily until their swarming, but you may hear them best in the evenings and mornings. Which music, as it cannot but please and delight those who listen to it, so must it be most sweet and pleasant to the young prince herself, to whom by means of is proclaimed a warrant, not only of her life, but of a kingdom also, both which otherwise she were to lose.

In this *Melissomelos,* or Bees' Madrigal[79], musicians may see the grounds of their art. First their moods, sometimes the triple or imperfect of the more, sometimes the double or imperfect of the left, then the tunes of the six notes, ut, re, mi, fa, sol, la, of which the queen sounds the first four, and the prince the other two, together with the doubling of fa-sol in two higher clefs, to make up the full octave, and lastly the six concords, an imperfect third, a perfect third, a diatessa[80], a sixt and a diapason. And if any man dislikes the harshness of the second and sevenths, which now and then hit among them, he shows himself to be an inexperienced artist, who knows not that as well as in music as economics, there must sometimes be discords, and that in either they have their laudable use, as serving to make sweet concords the sweeter. So that if music were lost, it might be found with the muses' birds.

---

79 Solmization is the practice of designating musical notes by syllable names. It was first introduced into European music by an 11th-century Italian monk, Guido of Arezzo. The six-note series, or hexachord, is used by Butler; it was later adapted in the seventeenth century to become a seven-note series.

80 diatessaron — late 14c. meaning an interval of a fourth;

The several parts of whose song comprises these mentioned notes, with pauses interposed (as I have at several times by a wind instrument, whose notes can neither rise nor fall, attentively observed) I thought good here to write down, that you may see in them all these particulars of their natural art. Only I cannot altogether warrant the conclusion, because in that confused noise, which the buzzing bees in the busy time of their departing do make, my dull hearing could not perfectly apprehend it, so that I was fain to make up that, as I could[81]. But I am sure, if I miss, I miss but a little.

In the morning before the swarm comes abroad, these ladies come down near the stool, and there they hold on their melody somewhat longer, singing sometimes above twenty notes together, and with shorter pauses.

At the very swarming time they descend to the stool, where answering one another in more earnest manner, with thicker and shriller notes, they begin to march along, thronging one another for haste, and buzzing with their wings in great jollity.

---

81 The first edition of FM (1609) had only a few notes to give the sound of bees buzzing; by the time of the 1623 edition, it had expanded to four pages. This may be because Butler's interests in music grew in the intervening years, when he was working towards the publication of 'The Principles of Musik' (1636). The Madrigal is printed in the custom of madrigal books of the time, so that four people holding the open book by its four corners or standing around it could each read their part. See The FM of Charles Butler, by *George Sarton* in *Isis: A Journal of the History of Science*, Vol 34, 1943

As soon as these gallant nymphs are aloft, they most nimbly bestir themselves, sporting and playing in and out as if they were dancing a country reel[82], in this manner waiting for the coming of their prince. Now when some two thirds or three quarters of the swarm has passed, the music stops and then comes forth this stately dame, Orithya, who, walking a turn or two before the door (on purpose, you would think, to be seen) she takes her leave, leaving but a small train to follow her, which rise after them after as fast as they can.

This decent order the great lords of the earth seem to have learned of this little lady, who in their country processions, their outings to Parliament, and other solemn proceedings, send the greatest and fairest part of their retinue before them, having behind but a small troop of necessary attendants, to guard their persons.

If the prime swarm is broken, the second will both call, and swarm the sooner; it may be the next day, and on that occasion perhaps a third also may arise, and sometimes a fourth.

But all within a fortnight after the prime swarm.

After the second swarm, I have heard a young lady-bee call, but the queen, not willing to part with any more of her company, did not answer, and the next day she (with seven more) were brought back dead.

Sometimes though the queen gives consent to a third or fourth, the bees seeing the stock little enough to live, show themselves loath to go, and then also there is no way with her, but one.

When the swarm is up, and busy in their dance, it is a common use, for want of other music, to play them a fit of mirth with a pan, kettle, basin, candlestick or other similar instrument, so to stop them, in truth, from flying away. Indeed, where other bee-folds are not far off, this use has a good use, for by this means the place and time of their rising is publicly notified, and so a swift and open claim laid to the swarm, that otherwise some false neighbour might challenge for his, which undoubtedly was the original cause of this custom. But the pretended reason is to me a mere fancy, although I know it to be as ancient as common…

If you see them begin to fly aloft (which is a token they would be gone) cast dust among them to make them come down. If they will not be stopped, but hastening on still, go beyond your bounds, the ancient law of Christendom permits you to pursue them wherever, for the recovery of your own.

---

82   Hey

But sometimes they fly so fast and so far before they pitch, that though you follow them never so fast, you must be content to leave them, happily to the happy finder. For when you have lost the sight and hearing of them, you have lost all right and property in them...[83]

Sometimes they will be provided of a house before they swarm, which some providers have found and viewed, and dressed in preparation for their coming, such as either a hollow tree or an empty hive, and then will they go away immediately, and by no means settle until they come thither. Unto which place they will fly, not, as at other times, uncertainly this way and that way, but as directly as they can guess.

A poor woman having taken a poor swarm to keep for half, by New Year's tide lost her own part and her partner's, and being careless of the hive when the bees were dead, she let it stand abroad until she had forgotten it. The next summer, coming into her garden, she found some bees passing to and from her hive, which were then busy in cleaning and dressing it. She, wisely fearing that the bees came to carry away the wax that was left, bade her daughter take the hive and carry it in. The wench, too involved in her play happily forgot her mother's command, and by that means, the hive stood still, until the unexpected swarm came, that afterwards accumulated in her garden. It is not therefore amiss to follow the counsel of Columella...[84]

When your swarm has made choice of an alighting place, you will quickly see it knit together in the form (if nothing else) of a cone, pineapple or cluster of grapes. As soon as it is settled, or at least as soon after as may be, hive them. For the longer they hang, the more loath they are to be put from the place, the more time they lose from their work, and the more in danger are they to be gone, either home again or quite away. For when they are once settled, they soon send forth spies, to search out an abiding place, who if they return with good news before swarming-time is past that day, they soon rise, and are gone; otherwise, they will stay until swarming-time the next day. But whenever the spies have spied, they return with all speed, and no sooner do they touch the cone or cluster, but they begin to shake their wings as the bees do that are chilled, which their neighbours, perceiving, copy, and so does this soft shivering pass as a watch word from one to another, until it comes to the innermost bees, which causes a great hollowness in the cone. When you see them do this, then may you bid them farewell, for soon

---

83   Butler appears to be alluding to texts like that written by the second century Roman jurist, Gaius: 'Among wild animals a distinction is to be drawn. In those of them that are half tamed, among which are mentioned deer, peacocks, pigeons, bees, property is not limited by strict detention, as in other wild animals... A migrating swarm of bees, accordingly, would only continue to belong to the owner of the hive as long as it continues in his sight and is easy to recapture, as it has no intention of returning.' The Institutes of Roman Law, 2,1,12-16

84   Lucius Junius Moderatus Columella, 4 – c. 70 BCE, *Res rustica*

they begin to unknit, and to be gone. And then though you hive them never so well, they will not abide.

When you see your swarm, first choose out a suitable hive, neither too big nor too little, but proportionate to the quantity and time of the swarm, so that the bees may fill it that year, or at the least to within a handful, which they may make up the next year in good time.

A swarm which is gathered before mid-Gemini, put into a hive that contains twice as much room as the swarm; a swarm at Cancer, into a hive that contains so much, and half so much, and for a swarm at mid-Cancer, a hive, that will hold it or little more, may suffice. Swarms between these times may be fitted proportionately into hives..

For example, a swarm of three gallons[85], or a good prime swarm before mid-Gemini, will need a hive of three pecks[86]; such a one at Cancer, a hive of five gallons[85]. Likewise a double-prime swarm coming early in the year is fitted with a bushel-hive[87], and all peck-swarms, and other single swarms after mid-Cancer, with the least, or half-bushel hive. But little and late swarms are rather to be united.

If this just proportion is not precisely kept, the bees may do well enough in a middle-sized hive, for being under hived, they will cast somewhat sooner, though it may be that fewer swarm, and being but a little over-hived, though they spend some time in supplying the former years' defects, they may swarm in good time, and become the fairer swarm. And indeed all swarms, whether bigger or less, by decreasing or increasing, do naturally draw towards this quantity.

But if the disproportion is great, it must be amended, whether you spy your error the same day, or afterwards.

If the same day, your remedy is to knock out the bees on the mantle[88] between two single rests, and to set a more suitable hive over them, but this is not to be done before the swarming hours are past, lest some of the bees take amiss, and go home again. Otherwise you may set the hive in a brake[89] with its bottom upward, and the fitter hive upon it.

If afterwards you see by the bees lying out, that they are under hived your remedy is to raise the hive with a skirt, or bolsters, as much as will let them in. If at Virgo you see, by their not filling the hive, that they are over-hived, your remedy is then to cut off the skirt as far as the combs, or near to them.

---

85   Gawn: a gallon
86   Peck: one peck is equivalent to two gallons
87   Bushel: one bushel is four pecks
88   Mantle: covering to the hive, as explained later by Butler: *'Your hive being fitted and dressed, you must also have in readiness a mantle, a rest, and a brush. The mantle may be a sheet, or half-sheet, or other linen cloth, an all square at the least.'*
89   Sticks placed in the ground to secure the hive

But generally it is safer and more to your profit, to under-hive a swarm, than to over-hive it.

Your hive being fitted and dressed, you must also have in readiness a mantle, a rest, and a brush.

The mantle may be a sheet, or half-sheet, or other linen cloth, an all square at the least.

A rest is either single or double. The single rest is a prism or three square column, eighteen inches long, and three inches deep, having the upper edge full of nicks for the space of six inches at each end, and the middle space, of six inches, smooth. It will be safer for the bees, and lighter for carriage, if the length of ten inches in the middle of the bottom is cut away one inch high, abating the new edges, and the four inches at each end are hollowed in the middle of the bottom from end to end, at the same height of one inch, and so this will be the form of the side:

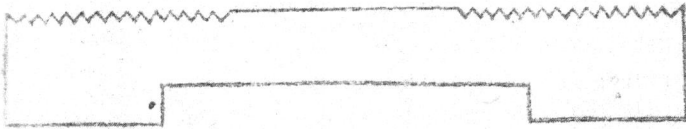

And this of the end. It is most suitably made of a quarter of a young tree:

On shelving or hanging ground, one single rest may serve, but if the ground is mostly level, it is better to use two, because the hive-skirt is set down upon them with less danger to the bees, than upon the ground or other flat thing. And these two rests are to be placed with the upper edges about nine inches apart, so that the hive standing upon them, may hang out over them some two or three inches.

In some cases two single rests are most convenient, but in most the double, which is also lighter for carriage, and more ready for use.

The double rest consists of two parts or sides, an inch thick, of the same length and depth with the single rest, having such upper edges so nicked at both ends, and the lower edges smooth, with ten inches of the middle cut away half an inch high, and then made sharp again; these two sides are to be fastened one to another, at the right distance of nine inches from edge to edge, with two rounds or braces entering into them three or four inches from the ends. The sides are suitably made of inch-board, or a cleft stick of willow or other wood.

The brush is a handful of rosemary, hyssop, fennel or other herb; of hazel, willow, plum-tree or other boughs, or rather of boughs with herbs, bound taper-wise together.

All things necessary thus prepared, let the person hiving – who must wear no offensive apparel – first drink of the best beer, and wet his hands and face with it, and then let him go about his business soberly and gently, taking good heed where he sets his foot, and how he handles them, for if he treads on a bee, or by any other means crushes one of them, they soon discover it, by the rank smell of the poisonous humour and will be so angry, that he shall have work enough to defend himself, unless he has on all his protective clothes, and being thus disquieted they will be the more difficult to hive. Moreover, the troubling of them does often make them rise and go home again; sometimes it breaks the swarm, causing parts to return, by which the rest are discouraged, being left insufficient; sometimes it disperses and spoils the whole swarm; it may also be the death of the queen, and then they will not continue to the next summer, however well provided. And experience has taught me[90], that few swarms, when greatly troubled in the hiving, will afterwards prosper. And therefore in any case, hive them as quietly and with as little trouble as you may.

The manner of hiving is many, by reason of the many and different circumstances of the alighting or pitching places, that it can hardly be taught by precepts, but is rather to be learned by use and experience, guided with reason and discretion. Nevertheless, for the help of novices, I will set down some special directions, which he that marks, may readily hive a swarm in most alighting places, and a little practice will fit him for any.

First therefore note that a swarm is to be hived by (1) shaking or (2) cutting the bough on which it hangs or by (3) wiping the bees down or (4) driving them up into the hive.

If your swarm alights on a bough, first spread the mantle under it and lay the rest or rests in the middle, with the ends toward two corners of the mantle.

Then if the swarm is so high, that you or some assistant may conveniently put the hive under it, having first removed the twigs round about, that stand in your way, shake the bees into the hive, and when you have set the hive right upon the rests, take up the two corners of the mantle at the ends of the rests, and pin them together on the top of the hive, to stop the bees running out suddenly, and then returning to the bough, shake it again, and turn it aside out of its place, or cover it with your body, or with some cloth, and then immediately loosen the corners of the mantle, and spread them again. When they begin to cease running into

---

90   'and experience has taught me' – a constant refrain throughout the *Feminine Monarchie*.

the hive, if you see them lie thick upon the mantle, shake them to the hive skirts, and the rest, as well upon the hive as the mantle, drive in gently with your brush. So shall you easily and quietly hive them. Otherwise having first taken away the twigs that may hinder you, cut off the bough or boughs (for sometimes they will hang on many), and if you doubt that some of them may fall in the cutting, let another helper support you with the ready hive, holding it directly under them. The bough being cut, lay the cone between two single rests, and set the hive over them. Or else put the cone first in the hive, and then set the hive down upon the rests.

But if they hang so near the ground, that you cannot conveniently put the hive under them, then placing the mantle and rests right under, shake them down, and setting the hive over them on the rests, take up the two corners of the mantle, and do as before.

And in case some of the swarm are first fallen to the ground, where they make no haste to rise again, then, placing a double rest without a mantle as carefully as you can, not killing any bees, either shake the rest down to them, and so set the hive over them all, or else set the hive over that part, and the rest, having cut the bough, lay besides the hive, and move them with your brush.

If they pitch upon a high tree, it is best not to shake them into the hive, but rather with a sharp knife cut the bough if you can conveniently, and either put it into the hive, and cover it with a mantle, or bring it down gently in your hand. But if you want a ladder or other means to bring it down, then let it down by a cord tied to some crook of the bough.

If they pitch on the body of a tree, or on some great arm, then set one side of the hive right over the bees, and with the brush drive them up by moving still the lower and wayward part. But if you have no means to fasten the hive by tying it above, or propping it beneath with prongs or the like, or if they are unwilling in this way to go into the hive, then parting them from the tree with a tight thread, wipe them down into the hive, and set them on the mantle and rest under the tree. If they are so high that you must climb for them, then cover them immediately with a mantle, and so carry them down. But look out should many rise again, in which case, leave well alone until they are knit together, and then sweep them in the same way into another empty hive, and put them with their fellows. If some of them still will rise up again, do not stop disturbing them, by wiping them off gently with your brush, by laying on mugwort, marjoram, wormwood, archangel, or other weeds or herbs, or by covering the place with a cloth, and after a while they will all go to their fellows in the hive.

But if they are so near the ground, that you cannot conveniently put the hive under them, then with a tight thread, sweep them down upon the ground, having first laid the rest either with or without the mantle, and set the hive over them.

And if they are of that distance from the ground, that you can set a stool close under them, then make fast one side of the mantle to the tree close under the bees, and the rest of the mantle lay on the stool with the rest; then, having suddenly swept down the bees onto the mantle, set the hive over them, and soon loosening that side of the mantle from the tree, lay it over the bees close to the hive.

If they alight on top of a stub, pollard, dead hedge, or the like, set one side of the hive over them, propping the other side with a prong or two, and drive them up as before.

If they alight in the middle or bottom of a dead hedge, your best way is gently to release them from the hedge until you come to them, otherwise you must violently knock the hedge on the other side, so forcing the bees into the hive, and then setting them down, trouble the place as before. But then be sure to be troubled yourself, for it is hard so to get them from such a position.

If they alight on some hollow side of a stub or tree, which they will be loath to leave, beware in any case you wet them not, for that does not only drown many, but also makes the rest more eagerly keep the place, because some through the wet cannot fly away, and their fellows finding them there will still resort to them. But when you have moved them by other means as much as you may, put some mortar or cloom into the hollow place, moving it forward by little and little, so that you bury none of the bees, until you have spread it over the place and then will they forsake that, and take some other part of the tree or stub, where you may more easily hive them.

When they fly into a hollow tree, so that by none of the earlier means can you hive them, then must you remove them by some offensive smoke, and make them choose a new alighting place, which is to be done in this way. If the bees lie about the hole where they went in (as they will do if they can) then bore a hole above them; if beneath, beneath them, but be sure that the upper hole is wide enough – rather than fail, make two or three with a two inch auger, or with a hatchet, one as great. Then fire a piece of match, or for want of a match, take a little hay, or another thing that will smoke moderately, and not flame, and put it into the tree beneath them, and you shall see them fly forth above for their lives, and soon pitch in some place where you may hive them. But this is to be done the same or the next day at the longest, for afterwards they will tolerate the stifling smoke and lose their lives rather than leave their goods.

If a swarm by reason of the coldness of the air, and roughness of the wind is not able to get away, and tries to alight on any other hive, quickly cover the hive close with a mantle, lest the bees entering it are pitifully murdered.

But in all types of hivings this one rule is general. The swarm must be continually kept together, for if at that time part of the swarm is parted from the company for the space of half an hour or less, afterwards when they find them, and would return to them, they are treated as strangers and robbers; as fast as they come they are beaten and killed. And those that escaping from there go back to their old home, find no better treatment, and those few that escape thence, desperately run into any other hives, and so leap out of the frying pan into the fire. And therefore when the swarm is hived, if you see part of it begin to gather together by themselves, remove them as speedily as you can, that they may go to their fellows in time.

And if you can choose, always set the swarm in the morning sun, and as near the alighting place as can be, which if some inconvenience will not allow you to do so, yet set it within the length of a perch, or at the least within sight and hearing, and then (lest those which are left at the alighting place, by losing their company awhile, lose their lives also) first trouble them by the means mentioned and then cause some of the hived part to arise by shaking them off the bough and by wiping them down that are on the outside of the hive. Which, when they are up, will make such a noise, that their fellows may easily find them. And if any bees are yet loitering by chance and likely to be set upon when they come to the hive, sprinkle the mantle, the hive and the bees with a little strong drink, and you shall part the fray.

And if any man marvels why those of the same swarm should so soon become strange one to another, seeing that bees of one hive are pent up for a whole day in another, and are yet welcome to their fellows at the last, I can give no other reason but this, that they knowing a swarm may part, and so each part becomes a several company, they deem these to be such by their long absence. And if you ask why they should find such harsh treatment in their old home from whence they came, it is because they went away with a leader of their own, and so became a company of several. And therefore if the leader stays away, those that come back (unless they come immediately) are treated as strangers, but if part of them have brought her home again, the rest safely return afterwards, either that evening, or the next morning.

If the swarm parts, as sometimes it will, and settles in different places nearby so that the bees can see each other, leave the greatest part alone, especially if it is suitable to hive, and disturb the other part in their settling by shaking, gentle rubbing with weeds and spitting and blowing in the place, that they may go to their fellows[91]. If they are settled and hang on a bough, cut the bough and bring them to them. If they are settled in some other place, then put them in a hive

---

91 Butler writes as a practical countryman in these pages, reinforcing his title page claim that the FM is *'written out of experience'*

without spleets[92], and if they are within a perch of the other part, move them both, one towards another, little by little until they are close together. After they have stood so for about half an hour, lift up the spleeted hive from its mantle and rest, and shake the bees out of the unspleeted hive upon the same; you must first knock the hive down, and then immediately clap it twice or thrice between your hands. This done, sprinkle both parts with good drink, and then without any delay set the spleeted hive over them, and they will immediately go up into it. But lay the unspleeted hive alongside close by, not where it stood, but on the other side, and those that remain in it will follow their fellows. But if the parts are further apart than a perch, then put them together the same night, as if they were two swarms.

In the same manner, when you have little swarms under the quantity of a peck, especially after Cancer is well advanced, put two or three of them together, whether they rise the same day or on different days.

For being thus united they will labour cheerfully, gather a store of wealth, and stoutly defend themselves against all enemies, whereas, if they were kept apart, they would surely perish at the next robbing occasion, or in winter, while if they remain alive, they would be of little good to you. And therefore if two swarms rising at the same time weld and knit together, (as they will do easily, if they are within hearing one of another) never trouble yourself to part them, nor be sorry for the chance. For those two being all one, are better than three such that are alone. Indeed sometimes it turns out that they fall out, and fight at the first, but that is because they are yet different companies under different commanders. For as soon as the inferior one is taken away, there remains one supreme monarch over all; the strife immediately ceases, and they are thenceforth linked in perpetual peace and unity together. Therefore they are little acquainted with the nature of these sagacious creatures, that draw their similarities from them, to cross that rich, mighty, renowned, thrice happy union, under one prudent, potent, peaceful, thrice noble sovereignty[93].

The way to unite two swarms is this. In the evening some two or three hours after sunset, or when it waxes dark, having spread a mantle on the ground, near to the stool, where this united swarm shall stand, and set a pair of rests in the middle of it; knock down the remover {hive} onto the rests, and then lifting up the hive a little, and clapping it between your hands to get out the bees that stick in it, lay it down on its side warily by the bees, and set the receiver on the rests over them, and they will begin immediately to ascend. If those that remain

---

92 See Glossary

93 The tensions in society which led to the outbreak of the Civil War in 1642 were very much present when Butler published the second edition of the FM in 1623. His appreciation of 'perpetual peace and unity' and the importance of 'the union' held together by a 'noble sovereignty' pointedly refers to the politics and society of his time as much as to the bees.

do not run out to their company, of their own accord, clap the place where they are gathered, and force them out, and lay down again the hive so that the small remnant may follow their fellows; if you spy any clustering by themselves, or straggling from the rests, guide them there. And when they are all in, either that night, or sometime in the morning, cloom the hive to its stool.

Otherwise about ten o'clock, or as soon as it is dark, set the remover in a brake with its bottom upwards, and the receiver onto it, binding them about the skirts with a long towel or two napkins sewed or pinned together, and so let them stand until the morning, and then set the receiver on its stool. In this way I united two swarms without the death of a single bee, apart from her that must not be saved[94].

If yet there are not enough bees in the hive, you may in a similar manner add another swarm to them.

In the uniting of swarms, two special inconveniences are to be avoided. The first is that being united, they should not exceed the natural quantity of a swarm, for if they do, though they are amicable and will gather, and grow fat, and cast the next year a fair swarm, yet will they never reach this size again, or scarcely swarm any more in that vast space. The second thing to avoid is that they do not fight, and destroy one another. To which two inconveniences is a third, that the swarms that unite themselves, if they are not aided, become highly disagreeable. To prevent the fruitless running together of more bees than are needed, which is the first inconvenience, when you see a sufficient fair swarm abroad, have an eye to the rest of your stocks. If you see another swarm about to rise, stop it by immediately shutting the door with a napkin, apron, or other such cloth, until the first swarm is settled. If then another one rising, draws near to it (as it will easily do if it can find the other), cover it quickly with a mantle until that is settled. If now being hived, another one presses into it, then before many have entered (that you may be sure not to have the queen) carry away the hive with the swarm about two perches off, and set an empty prepared hive in its place for that other swarm.

If none of these things are done, but swarms run together in greater quantity than a good hive can contain, then rear the hive with bolsters high enough to let them all in, which, when they have once swarmed, the next harvest time take them away.

In a fair afternoon, about four o'clock, pick away all the cloom between the hive and the stool, and in the morning, at the break of the day, lay the hive along with the edges of the combs up and down, on a mantle spread on the ground, and there pare off the combs' ends evenly with the skirts, and so set it again on

---

94   i.e one of the two queens.

the stool on moveable door posts, and a thin bolster behind, and cloom up the hive as closely as may be.

Concerning the other inconvenience, know this, that although two strange swarms, with their several queens, never meet in one hive without discontent (which they express by running to and fro outside, and making a tumultuous noise within, after which they sometimes fall to fighting and killing) yet commonly this strife is soon at an end. For the first queen having got the right of the whole space by the possession of the capital or superior part, where she sits tight with her guard about her; the inferior by a common consent, is straightaway despatched, and so they all become fellows and friends under one sovereign. And therefore when swarms are united by you, be sure that the bees in the receiver are not thrown down among the others, lest the superior queen comes down with them, and so you create more strife than is needed.

But the danger is when two princes with their equal sized colonies happen to be equally advanced in the hive, and therefore neither yields to the other, but fight it out on both sides with equal hope of victory. When this happens - which is very seldom - the controversy is undecided, and the conflict is likely to be perilous, or rather pernicious, if it is not prevented. In this case you have no other way but the next morning, if they still fight, to cast them all out of the hive, and so will they either knit apart, or return to their old stocks, from which another time they may swarm more luckily. The sixth and twentieth of June, 1621[95], I had two fair swarms up at once, which going together overfilled a good hive, where neither of them yielding their queen to the other, the fight continued two full days and two nights, even from Thursday noon until Saturday in the afternoon, during which such havoc was caused, that the better part of these brave soldiers (a mournful spectacle) lay some dead, some half-dead sprawling on the ground. At the last it was my good fortune to spy one of these queens at the hive skirts in a cluster, which taking up, now, said I to one that stood by me, here is she for whose sake all this slaughter was made. About an hour after my son found the other dead on the ground. When they had thus mercilessly murdered both the queens, and the better part of the swarms, those that escaped rose all out of the hive, and went into another swarm which stood behind them, of which, because they brought no ruler with them, they were quietly received.

Sometimes a swarm when abroad, and gathered together in the hive cone, will not continue, but will return home again, the causes of which are several: windy, wet, or cloudy weather; not finding a suitable alighting-place; trouble in

---

95 Butler gives very few specific dates in his book but here on 26th June, 1621 he refers not only to an incident he has recorded, but also, in passing, to his family. The *'one that stood by me'* is quite possibly his wife; he also refers directly to one of his three sons, who might, in 1621, have been his youngest, Richard. His wife died in 1628.

hiving; the hot position of the hive lacking protection, or the missing of their prince. And this, especially when it is during a plentiful season, they are then ready to return for little or no reason, and reluctant to go abroad, even in the safest weather. I observed once, that the prince being scarcely ready, fell down from the stool unable to recover her flight, upon which the swarm returned. She was put into the hive, and the next day the swarm rose again and settled, but the prince happened to fall beside the cone. The swarm being knit together and missing her, began to unknit, and go, which I perceiving immediately hived them, but they being still discontented, ran up and down the hive, with a murmuring noise both without and within. Directly after that I had watched a handful of bees hanging on a nettle on the ground, among which was the prince. When I had cut off the nettle, and set it by the rest under the hives' skirt, immediately the knot unbinding, I saw the lost prince with her long train stately walking into the hive. As soon as she had entered, these malcontents began to stand still and buzz, joyfully shaking their wings, as they are prone to do when they are pleased, and so they quietly maintained the hive. To see the sudden alteration among them immediately on her approach, and how they could have notice of it all at once, the ones outside as well those within, would even make a man to wonder, if it were not for the fact that all they do is nothing else but a set of wonders.

Swarms that go home, do sometimes stay long before they rise again, and when they rise (especially if they were hived) they are likely to fly away, although I have known a swarm to rise four times in three days, and at the last to be quietly hived. If therefore you perceive the swarm returning, before many have entered the old stock, shut the door fast; if that will not do the trick, carry the old stock away, stool and all, and set the swarm immediately on a stool in its place.

And if any of them are going into other hives (as sometimes, where the hives stand near together or are many, some of them, especially the young nymphs that have not been abroad before, will do) cover them with mantles, for as many as enter will die, or barely escape.

If a swarm alights near the place where another was hived a day or two before, be sure to set it as far as conveniently you may, from the place where the former alighted and stood; the space of a perch or somewhat less may suffice, otherwise many of the first swarm resorting there, will go to the new swarm, and so be killed.

When your bees are hived, those that hang on the outside, drive in gently with your brush, and lay the corners of the mantle that are farthest from the rest, over the hive, with boughs also to shadow it, if the weather is hot. But if you find them unwilling to go in (as in extremely hot weather they will be, though they like the hive well enough) then strive not with them, but laying the corners of the mantle over the hive, as before, with boughs to shade it, there keep them until the heat has abated, and then drive them in, and if you think they cannot otherwise well endure that heat, cover the hive again with mantle and boughs. And so let it stand until it waxes dark, and all the bees have come home.

Then knitting the four corners of the mantle together, at the top of the hive, and binding the mantle about close to the middle of the hive with a small line, carry the swarm to its place. And after a while, taking away the mantle, set it upon its seat with the door toward the south, or rather south-west, and then leaving only a breathing place, for fear of stifling them, cloom it up close and put on a hackle and so let it stand until it is fair and warm the next day. For if the hive is left open, in the morning sometimes they will resort to their former standing and there abide, sometimes flying about, sometimes settling on the ground, where if the cold or wet takes them, many die. When you see the weather suits them, then hanging the mantle, or other white cloth on the hive, let them go. But they will the sooner leave the haunt of their hiving place, and fall to their work, if you show them their new standing by knocking them out together onto the stool, when the weather is warm.

All swarms, if the morrow is fair, will desire to be abroad early, and knowing their want, will bestir themselves more lustily in their labour than other bees. But if the foul weather keeps them in the first day, then are they much discouraged, so that the next day being indifferent, while other bees work hard, they will scarcely look out of the door, not daring to commit their hungry and thin bodies[96] to the cold air. And if they are completely kept in the second day also, then will they not stir (even though they die for it) until the weather is very pleasant. They may live five or six days in the hive without honey, but afterwards they begin to string down, hanging one at another's heels, which is a certain sign of death, if they are not soon relieved.

To prevent this evil, if the swarm alights in your garden within a perch of the seat that is appointed for it, set it there at the first, and so will they lose no time in loitering about the hiving-place. And if it alights further off (whether in your garden or another place where they may stand safely, especially if the weather is unkind or inconstant) leave them there until it mends; for those that are not removed, but keep still their first standing, because they are do not seek their way home, they fear the foul weather as little as the best. And therefore they do not need to be shut in the morning, as those that are removed, or to have any white cover over them for their direction.

---

96   The expression *'leere and thinne bodies'* suggests that Butler was not aware that a swarm, prior to departure, will feast on honey in readiness for the journey

The means to recover such a drooping swarm is this. The first sunshiny day, turn up the hive to the sun, that this heat may revive them, and sprinkle a little the sides of the hive, the spleets, and the bees with mead or honey-water; hold them so in the heat of the sun until you see many of them fly abroad. Then set down the hive gently on its seat again, and cover it not until it is thoroughly warm, and the bees play cheerfully, as at other hives.

Chapter 6
Of the bees' work

Nothing is more odious to the industrious nature of bees than sloth and idleness. While there is matter to work upon (unless they are hindered by unkind weather) their work never ceases; the old bees, which have spent their days in continual labour, will not at the last allow themselves any immunity or rest in their hives, as a recompense for their pains' past, but continue still their travail unto death. Indeed, in the three still months - Sagittarius, Capricorn and Aquarius - because then there is nothing to gather, they do not work (yet when a fair day or hour comes, as weary of rest, they will abroad, employing themselves in different and necessary offices), but so long as any good flowers grow, even from Pisces or a little before, up to Sagittarius and, some years, somewhat after (which is a full nine months) they lose no time but follow their business tooth and nail. This incessant labour while the time permits, which has three unique effects: (1) the working of wax, (2) the making of honey, and (3) the feeding of their young, the poet Virgil in a few words has elegantly expressed...

Their first work is the ground of the other two, the artificial cells serving both for boxes to lay their sweet treasure in, and for nests and nurseries to breed their young in. The matter for which they gather from flowers with their fangs, which, being kept soft with the heat of their little bodies, of the air, and of their hives, is wrought into combs. This work is so nimbly and closely done, that it can hardly be perceived, as Aristotle notes...[97] But Pliny[98] is willing to go a little beyond him, telling us a tale of a lantern hive made at Rome, through which, in truth, the bees' doings in the hive were described, and in another place he writes of another similar happening... But unless the bees were also transparent, as well as the hive, this cannot be the case, seeing they do always frequently compass the combs round about. A more likely way than that, were to have a moveable piece in one side of the hive, which when you have taken it away, you may see the drones and the honey bees walking together to and fro, and with their doubled heat hatching their young, but their work you cannot see, even though you remove and part the bees until the bare comb appears. But if your curiosity longs to see the manner of their curious and artificial building, the only way is this. In Gemini set up a last year's middling swarm two or three handfuls above the stool and then when most of the bees are abroad (but most suitably in the forenoon when they are quietest) you may behind the stool behold them working on the edges

---

97 *Histories*
98 *Natural History* chapter 16

of their combs, and having blown their liquid and soft wax out of their mouths (as the wasps do their worthless, drossy stuff, which you may see them gather from wooden fences[99] with their fangs and so carry it away) to fasten and fashion it with their fangs and forefeet.

How much wax they bring at once, appears by the new swarms, whose first work is spent chiefly in building combs, at which they are so earnest, that it falls out with them as it is in the proverb, *The more haste, the less speed.* For many of their burdens fall from them before they can fasten them to the combs. You may then see a great store of them on the stool by the skirts of the hive, similar to the white scales, which fall from young birds' feathers. And therefore some have imagined, that they also are scales which the young bees likewise shed from their wings. But if you put some of those parcels together with warm fingers, your doubt will soon cease.

The bees' combs are placed differently than that of the wasps, for the wasps hang theirs one under another, and the bees theirs one beside another, beginning in the top of the hive, at such distance that a bee may reach from one to another.

Their cells or little holes are made six square, according to the number of their feet, and of that length and width, that each of them may easily contain a bee. These are so artificially wrought and joined together that St Ambrose refers to them…[100]

But here their art is yet more exquisite, that whereas there are two courses of cells in the two sides of every comb, the cell-bottoms in these two sides are never opposite one to another, but each hexagonal bottom of one side answers to three third parts of the hexagonal bases of three contiguous cells on the other side, meeting all in one angle right in the centre of the opposite bottom, as in this form,

which is so artificial, as well for strength as beauty, that no young bee, even though the thin bottom of his cell should fail, can break through into a cell of the other side. He that sees this, does he not see a wonder?

Besides these ordinary combs, there is commonly one drone-comb in a hive, in which the young drones are bred, made for that purpose with wider cells… although in some hives part of the drone-combs are made out with nymph cells.

---

99   *Pales*
100   Hexameron, Book 5, Marine life and Birds

The drone-comb are no thicker than others, and yet the drones are longer than the small bees; they increase the length of his cells by covering them, not with a flat cover, as they do the rest, but with a deep hollow one like an old wives' fringed cap, which afterwards, when the drones are bred, they take away. And when those cells are empty of young drones, they fill them as they do the others with honey, and after swarming time, if they want upper cells for their honey, they will not tarry until their drones come forth themselves, but liking better their rooms than their company, they draw them out of their seminaries[101] before they are ripe.

But the queens' cells are built singly, everyone one by himself[102] and in different places of the hive: some above and some beneath, that, like other princes, she may for her delight move there at her pleasure. But for the most part, in the outsides of the combs, for although it is fit for princes to be near their chief cities, yet they do not love to be pestered in the middle of them. In fashion, they are round, which is the most perfect figure, as the six square is most suitable for comely joining many such buildings together. They are also larger than the rest; to show that subjects' houses should not match their sovereign's in greatness. In these palaces they breed their young princes...[103] The common people, finding them always in those stalls that die, take them for certain signs of death, and call them pipes or taps, and therefore when they see them in a stall that they take, they say, This was taken in good time, for it is piped, and therefore would have stood no longer. But seeing none are without, no not even the youngest swarms, ordinary reason might teach them to forgo that foolish idea.

The combs have successively sundry colours: white, yellow, brown, black. Their first colour, white, by the end of summer is turned to a light yellow. Those that are taken and tried this first year are called virgin-wax, but the whiter the purer, and the rest are ordinary. The second summer this light yellow is changed to a more sombre one. The third summer, the dull yellow changes into brown, which afterwards, as they wax old and corrupt, alter again into a blackish and dirty colour, but these being melted[104] will return to yellow.

The time when bees gather wax is only between Taurus and Virgo (unless Aries is mild and warm) for then they may begin in that month.

But honey they gather all the year, except only in those three still months, when the weather keeps in both bees and flowers. And it is of two sorts: the one pure and liquid, which is called nectar; the other thick and solid, which we may

---

101 'Seminaries' from Latin seminarium, 'seed-bed', and later used for places for training priests, which suits Butler's almost mystical reverence for the bees.
102 *'Himself'* is commonly used for both male and female; Butler uses *'her'* of the queen in the next lines
103 Pliny, *Natural History*
104 *'Tried'*

for a similar reason term *Ambrosia*. For both serve for the food of these divine creatures. The thick honey is gathered by their fangs, from there it is conveyed by the fore-legs to the thighs of the hind legs…and so nimbly, that unless you have a quick eye, you can scarcely perceive it.

When they have brought these burdens home, they unload them into the dry cells for the young to feed on, which are not yet able to fly abroad. And in the beginning and ending of the year, observe what they save when the weather is fair: this they lay up for themselves against a rainy day. Which, while it is good, they will feed on, to save their nectar as much as may be. But this kind of honey is similar to fresh fish: it must not be kept long. For if being laid up in the cells, by reason of plenty that comes in fresh and fresh, it lies unspent; after a while it corrupts, and from being sweet it becomes the sourest and the most unsavoury of all things both to taste and smell, which then becomes decayed honey and a wasted residue. Where there is any store of this stuff, it does so offend the bees, that often it makes them forsake all. Most of them will that year go out in swarms, and those few that are left will never prosper.

Concerning this leg-honey or crude honey there is a general error. For, without scruple or doubt, men count it and call it wax (as did some also in time of old, whose opinion Aristotle conveys…) But against that view, (as I shall show you) there is both sense and reason[105].

If you put it to your tongue, it has the taste of honey, which wax has not. If you feel it between your warm fingers, it breaks[106] apart, whereas wax sticks fast together. If you put it to the fire, it melts not, as wax does. And whereas wax is all of one colour, white at the first, even as those little fallings of the new swarms (which is wax indeed) this leg-honey is of different colours, white, black, yellow, green, red, tawny, orange, murrey, and of various middling colours. Therefore common sense suggests it is not wax.

The reasons are two. The first is, because when they gather an abundance of this stuff, the bees do not acquire any more wax. The other is because when they make most wax, they gather none of this.

For proof of the first, all the bees between Virgo and Taurus gather an abundance of it, and yet their combs are not during this time in the least enlarged. Also one of those old stalls that are full of combs carry more of this matter all the summer long than many swarms, and yet they have no more wax at the end of the year than at the beginning.

For proof of the other, the new swarms in one week, if the weather allows them, will have half filled their hives with combs, and yet in all this space you shall

---

105 Butler continually subjects the classic authorities to the twin tests of reason and experience, in keeping with the growing scientific character of the seventeenth century.
106 *Muttereth*

scarcely see one carry any of this. If you would know the reason why the stocks gather so much, and the new swarms so little, it is because the stocks have young bees which they feed with it, and the new swarms have none. And if any foolish bee does carry in Ambrosia, it is put in a dry cell where it turns to decayed honey, as I have seen within a fortnight after the hiving.

And this, though it seems new, yet was it known many ages ago. Pliny wrote of it… and before him Aristotle…

The nectar or liquid honey the bees gather with their tongues, whence they let it down into their bottles, which are in them and are similar to bladders; each of them will hold a drop at once. You may see their little bellies strut with it all. Men think, because they see nothing on their legs, that they come in empty, when they are more fully and more heavily laden than the others. These bottles, as soon as they come home, they empty into their combs…This nectar, being as clear as crystal at the first and liquid as water, when it is two or three years old, becomes white and hard…While it continues as liquid and will run of itself, it is called live-honey; when it has turned white and hard (even like sugar) it is called corn-honey, or stone-honey.

And the hive-honey is of two sorts, that which is gathered by a swarm, clear and crystalline at the first, laid up in virgin wax, and taken the same year, is the true virgin-honey; the other, which is yellow and thicker, gathered by an old stall, and therefore kept in more corrupt cells with dross and coarser, is called ordinary.

The first flow of which (especially in a plentiful year of nectar-dews) runs completely by itself, and is a kind of virgin-honey and little inferior to the true.

Nectar, whether it is ordinary or virgin-honey, is either finer or coarser, according to the soil where it is gathered. For the best countries, which yield the best wheat and the best wool, yield also the best honey. And therefore the woodlands of Hampshire[107] have better honey than the heath, and the champion or field country, better than the woodlands. The reason is, because whether the flowers are most fragrant and virtuous, as well of the fields as gardens, in the purest and sweetest air, there the honey-dews, which are extracted from them, are most fine and pure.

When the cells are full, they close them up with little films of wax, which they will not break until winter and hunger drives them to it. And thus do they all the summer, descending lower and lower from one cell to another, until Virgo, after which time they lay up no more in store. For honey then waxes scarce abroad, and from thenceforth they can gather no more wax to shut it in. As for that which they purchase by fight and foraging, it does them little good. For the most part they soon spend, and if they save any, they half fill a few cells with it, which being

---

107 Butler lived nearly all of his adult life in Hampshire, and it was the English county he knew intimately.

uncovered, either themselves or some other thieves quickly devour, according to the Proverb, 'Evil gotten goods are soon spent.' [108]

This nectar and ambrosia, together with those sweet and wholesome vessels that contain them, are gathered from an infinite variety of herbs, flowers and trees which God in his provident bounty has ordained to succeed one another. So that from Pisces to Sagittarius, there is never a shortage of some plants or other, containing these sweets, which the bees skilfully draw from them, without any hurt to the fruits.

The dent-delion, or after the French pronunciation, dandelion, may well be called *apiastrum* or *melissophyllon*. For the bees gather upon it almost all the year. The daisy and yellowcrea are next for continuance, but nothing nearly so much regarded.

The winter gillyflower and the hazel are the first. For they spring in Pisces, and sometimes before. After them the daisy and the herb bear's foot[109], the violet etc.

In Aries besides those earlier named, the box, the willow palm, both green yielding nectar, and yellow yielding ambrosia, daffodil, the March lily, blackthorn etc.

In Taurus sloe-tree, plum tree, gooseberry not blown, and blown, cherry, pear, cockbell, which is a wood flower. About the middle of this month the chief plants begin to flourish in great abundance, such as apple, crab, barberry, crowpicks, charlock, rosemary etc. But especially the plentiful vetch and maple. They gather on the flower of the maple a whole month together, and a certain amount on the flower of the vetch when it's time has come, but the greatest store of honey is drawn out of the black spot of the little picked leaf of the vetch, which grows on each side of the two or three uppermost joints. These they ply continually. I never saw vetches, however far they were from hives, that for three months together (if the weather served) were not full of bees.

In Gemini, the first month of fruitful summer, besides those prime plants, vetch and maple (which now are in their prime) and the rest forenamed; beans also, which with their flowers have also black spotted leaves like vetches, on which sometimes they gather, archangel, barberry, fumitory, ribwort, a kind of plantain, holm or holly, hawthorn, elder, red honeysuckle, redweed, white honeysuckle, which they like much better than the red.

In Cancer, with the forenamed, the blossom of the vetch, as well as the leaf, benet, mallows, the sovereign thyme, which yields only nectar, and therefore he was deceived that said 'crura thymo plena' [110]. Thyme, for the time it lasts, yields the most and best honey, and therefore in old times was accounted chief...

---

108 The Proverbs of John Heywood, 1546
109 *Helleborus foetidus*
110 'Legs full of honey'

Himettus in Greece and Hybla in Sicily were so famous for bees and honey, because there grew such a store of thyme…The knapweed flourishes about the middle of this month, and the blackberry about a week after, both of which, as sweet and plentiful, the bees much haunt.

But the greatest supply of the purest nectar comes from above, which Almighty God does miraculously distil out of the air…and has ordained the oak, among all the trees of the wood, to receive and keep the same upon its smooth and solid leaves…until either the bee's tongue, or the heat of the sun has drawn it away. When there is a honey dew, you may perceive by the bees, as if they smelled it by the sweetness of the air, they immediately issue out of their hives, in great haste following one another, and their old haunts, search and seek after the oak, which for that time shall have more of their custom, than all the plants of the earth. Sometimes the maple and hazel take part with the oak, but little and seldom. While the honey dew lasts, they are exceedingly earnest, plying their business like men in harvest: you may see them so thickly at the hive door passing to and fro, that often they throw down one another for haste.

What this 'mel roscidum' [111] should be, Pliny seems much to doubt…But if conjectures might be allowed, I would rather judge it to be the very quintessence of all the sweetness of the earth (which at that time is most plentiful) drawn up, as other dews, in vapours into the third region of the air, by the exceeding and continual heat of the sun, and there hardened and condensed by the nightly cold into this most sweet and sovereign nectar, and then does it descend to the earth in a dew or small drizzling rain, causing one writer to comment on it…[112]. His opinion is the more probable for these reasons. Firstly, because when the year is backward in its fruits, the honey dews are also backward, coming only at such time as the flowers have the most solid and best juice. Before, when the juice is weak and waterish, and afterwards, when it is dried and wasted, they are not. Secondly, because in hotter and more southerly climates, where the fruits are more forward, the honey dews also are more timely, as in Italy before Gemini… And thirdly, because the countries that have store of the best and sweetest flowers, always have the best honey.

The hotter and drier the summer is, the greater and more frequent are the honey dews; cold and wet weather is unkind for them; much rain at any time, coming from a higher region, washes away that which is already elevated (so that there can be no more until another fit of hot and dry weather) and in the end it quite dissolves them away.

---

111 'honey dew'
112 Galen, c.130 -210 CE

The time in which these honey dews fall, is usually between the first and last days of this month, although the continuance of hot and dry weather may cause them to come somewhat earlier, or last somewhat longer, even until mid-Leo or after. They may happen at any time of the day, but for the most part in the morning before it is light…And then shall you have the bees up in a morning as soon as they can see, making such a shrill noise where they go, that, as merry gossips when they meet, a man may hear them farther than see them.

In Leo, vetches, mallows, thyme, knapweed, blackberry, white honeysuckle, redweed, thistle, melon etc.

Now also do they gather on the lavender, if their hasty dames do not gather it from them before it is ready.

In Virgo, knapweed, blackberry, redwood, dandelion, mallows, borage, etc and the ample heath, which yields honey similarly as it does wool.

In Libra, dandelion, heath, Iuie [113] etc.

In Scorpio dandelion, Iuie, archangel etc.

And in this great variety this is strange, that where they begin they will make an end, and not meddle with any flower of other sorts, until they have their load… Insomuch that those which begin with the flower of the vetch will not once touch the rich spotted leaf of the same, before they have been at home. Although when they come to a flower that yields both nectar and ambrosia they will use sometimes the tongue and sometimes the fangs, and gather them back.

But this may seem more strange and wonderful, that out of the most stinking and poisonous weeds, such as redwood, marges [114], henbane and the like, they gather most sweet and wholesome honey, and yet regard not some of the best and sweetest herbs and flowers, such as the rose, the primrose, clove gillyflowers, wheat, barley, peas etc.

What store of wax and honey a stall may gather is uncertain, some having more, some less, according to the number of the bees, the size of the hive, and the plentifulness of the years. With us[115] it is counted a good stall that yields two or three gallons[116] of honey[117], although in a tree there have been found more than seven or eight. But in other northern countries we read of far greater quantities. Pliny affirms that there was seen in Germany a honey comb eight feet long. And Paulus Iovius[118], that in Moscovia there are found in the woods and wildernesses great lakes of honey, which the bees have forsaken, in the hollow

---

113 'Iuie' - possibly a reference to Tradescantia which was introduced into Europe as an ornamental plant in the seventeenth century.

114 Stinking chamomile

115 Domestic reference to Butler's family enterprise in honey- making.

116 *Gawnes*

117 *'Pulse'* – perhaps the unstrained mix of wax, honey and comb

118 Paulus Jovius, also called Paolo Giovio (1483 – 1552) Italian historian, Bishop of Nocera.

trunks of marvellous huge trees. Insomuch that honey and wax are the most certain commodities of that country. Where, by that occasion, he sets down this story, reported by Demetrius, a Moscovite Ambassador sent to Rome. A neighbour of mine (says he) searching in the woods for honey, slipped down into a great hollow tree, and there sank into a lake of honey up to the breast, where, when he had stuck fast two days, calling and crying out in vain for help, because nobody in the meanwhile came near that solitary place, at length when he was out of all hope of life, he was strangely delivered by the means of a great bear, which coming thither about the same business that he did, and smelling the honey stirred with his striving, clambered up to the top of the tree, and thence began to let himself down backward into it. The man thinking to himself, and knowing that the worst was but death, which in that place he was sure of, clasped the bear fast with both his hands about the loins, and in addition made an outcry as loud as he could. The bear being suddenly frightened, what with the handling and what with the noise, made up again with all speed possible; the man held and the bear pulled, until with brute force he had drawn *the horse out of the mire* [119], and then, being let go, away he trots, more afraid than hurt, leaving the smeared young man in a joyful fear.

The bees' earnest and hot labour, and the drought of the air, together with their heated constitution [120], which their very hue betrays, causes them to desire cold water a great deal. Some think it serves only to feed their young ones … and that not without reason, seeing that ambrosia their daily food is hot and dry, and indeed when the drones are done away, and breeding is ended, the bees are nowhere near so frequent at the watering places. But Columella thinks the use of water is more general… to whom the poet, in the place first cited in this chapter, seems to assent, making water and flowers the common matter of their three works.

The watering place should not be far from your garden, in the next side of a pond or brook, made shelved, not very steep, in the manner of a ford, and defended from beasts, geese, ducks, and such like, and especially young ducklings.

But because in the cold windy weather of the spring (at which time of the year the bees have the most use of water) these watering places of ponds and brooks are dangerous, where you may then see many thrown down drowned, and others, that escape drowning, to be so chilled, when they have filled themselves with cold water, that they are not able to endure the wind, but fail and fall by the way, therefore it is necessary to have troughs in your gardens, made for the purpose, where the bees may both sooner and more safely fetch their water.

---

119 Popular proverbial expression: *'dun out of the mire'*
120 *Choleric complexion*

For the form and size of a trough, let its hollowness be two feet in length, seven or eight inches in breadth, and four in depth; the bottom four inches thick, the ends six or seven, and the sides half so much. Moreover, let the hollowness be divided into four equal parts, by one partition of inch-board, in the middle from side to side, and by two partitions of half-inch-board, from each end to the middle partition, after this fashion:

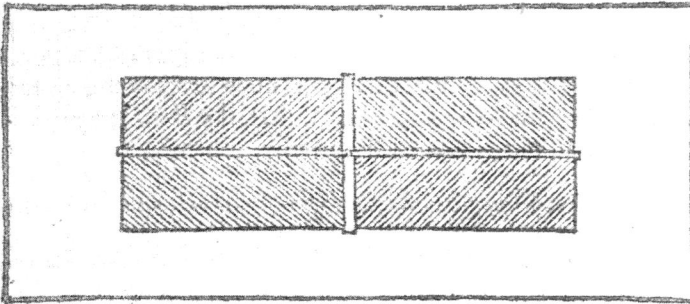

And to keep the bees from the danger of drowning, to which they are very liable (for if they but touch the water with their wings, they cannot rise from it) let each quarter of the trough have its cover, in thickness about half an inch, in breadth and length fitting to its quarter, but so, that without hindrance it may rise and fall with the water.

The matter of this cover must be cork, which must as well have open spaces for the water to take air, as places for the bees to light on, lest it being covered too closely, does spoil and become unsavoury. It is best to divide each cover into two equal parts, and in the edges on both sides to cut little nicks. And so this may be the form of it.

Other fashions both of their troughs and of their covers, may be devised, but these have seemed to me in all respects most fitting.

A new trough thus framed and fitted, is to be seasoned before it is used, by often scalding it, and changing the foul water, until, having stood a day or two, it remains clear, and without a glittering slime; afterwards, the older and more earthy it is, the better they like it.

The trough being seasoned, set in some convenient place, about a perch from the bees, having a movable plank or the like, to defend it from cold rough winds in the spring, and from the sun when it is hot. At which time keep the trough full, lest the water is soon overheated, and in cold weather let the water remain shallow, that the bees may drink safely below, out of the chilling wind.

You may also make good troughs of freestone, with wooden partitions let into the stone, but they are more apt to chill the bees in cold weather, until they become mossy.

Sometimes the bees will lie sucking at nearby standing water [121], puddles and mire in the streets, where many are trodden under foot by men and beasts. See therefore that such places are kept clean and dry.

After a shower they draw water for the most part in your garden from the bare earth, the grass, and wherever they find it wet from above. In the chief breeding months - Aries, Taurus and Gemini - when the cold rain or wind has kept them in for some part of the day, they will lie so thickly on the ground, if you have much stock of bees, that you can scarcely tread beside them. At such time therefore let no heedless stranger come among them.

121 *Plashes*

Chapter 7
Of the Bees' Enemies

The good bee, as other good things, has many enemies, from which she needs your help to defend her, especially, (1) the mouse, (2) the woodpecker, (3) the titmouse, (4) the swallow, (5) the hornet, (6) the wasp, (7) the moth, (8) the snail, (9) the ant [122]), (10) the spider, (11) the toad, (12) the frog, (13) the bee, (14) the weather.

The mouse, whether he is of the field or the house, is a dangerous enemy. For if he gets into a hive, he tears down the combs, makes havoc of the honey, and so starves the bees. Some enter by the door, or by some open place in the skirts of the hive; some gnaw a hole through the top of the hive, where they know the honey lies; some keep their old homes, and come to the hive only for their food; some make their nests between the hackle and the hive, that they may the sooner and the safer come to the honey at their pleasure. For remedy, first you must see that your hives, whether they are of straw or wicker, are closely made and secure. For if the straw is loose and soft, they will more easily gnaw their way through, and if the wicker is thin, when they have torn down the cloom, they will creep in between the twigs. Next see that the hives are daubed close round about by the skirts, that there is no entering but by the door, which in Taurus, when the bees come down to watch, and throughout all the summer, they will keep well enough both day and night, but all the winter, at which time the mice plunder the most, it must be made so narrow, that they cannot get in. Also it is necessary for you to remove all things around your hives, that may hide and harbour them, for they will fear to come and go when in view, lest the cat meets with them by the way. In addition, it is good now and then, in dry and warm days, to take off the hackles, as well for this as for other causes. Those that nestle on the top of the hive, when the hackle is taken off, will sit still, bewildered, for so long, that you may be sure to crush them against the hive with your hand. Lastly, you do well to set baited traps in their way, that so they will reach only a short distance into the hive.

The woodpecker or *yippingale,* if he finds any hole in the hive against the honey, does with his long round tongue draw it out, but he does more harm to wood bees than garden bees.

Of titmice, there are three sorts. The great titmouse (which, because of his sooty head and breast some call a colemouse) is a very harmful bird. For although sometimes he seems content with dead bees, yet is he a great devourer of the living also. In winter he takes them at the hive as they come forth; when the cold makes them keep in, he will stand at the door, and there will never leave knocking until one comes to see who is there, and then suddenly catching her, away he flies with her, and when he has eaten her, he comes again for more: eight or nine will scarcely serve his turn at once. If the door is shut so that none can come forth, he labours to remove the bar; if that is too heavy, he falls to moving about the door for a new way, and when these methods cannot get them out, some have the skill to break the walls of the daubed hives above, over against the place where they lie, and there they are sure to have their intention. But in the spring, when the bees come to the palm, he stands there watching for them, and while they are busy at their work he devours many. The little russet one in the winter feeds only on dead bees, but in the spring he will take part with the great one. The little green titmouse I cannot accuse, unless it be for eating a few dead bees, and that is seldom, except in some hungry time.

The swallow takes them as they fly...but I am persuaded she does much less harm than the titmouse, although she has a worse name. The long-winged hawk makes the fairer flight, but the short-winged is the kitchen-hawk. These birds are therefore not to be tolerated...

Let boys destroy their nests in summer, and catch the titmouse in winter, with traps baited with dead bees, oats or tallow. Aristotle joins the wasp, the little titmouse, the swallow and the great titmouse together... {*in his list of enemies*}.

The hornet also devours bees, being much too strong for them, that they can make no resistance, to which the poet [123] refers...

The hornet's manner is to fly about before the hive, until she has spied her prey settled at the door, and then suddenly she takes it in her feet, and flies away with it, as a kite with a chick.

In destroying the hornets you must be wary, for one stinging often causes a fever, and less than thirty, as some say, will kill a man...

The wasp does much more hurt than the hornet. For the hornet now and then kills a bee, but the wasp wastes the honey, by which many whole stalls do perish. For besides the harm that she does herself, she often sees the robber at work, who, when the wasp has begun, will be ready to take part with her, and then all goes to wrack. A wasp is by nature harder and stronger than a bee, especially in Libra, inasmuch that often she breaks from two or three of them, though they have all hold of her at once, and perhaps kills one of them out of hand. At Cancer, or if the spring is hot and dry, in the later part of the former month, the wasp begins to breed within a month after she first appears; and in a while she begins to feed upon dead and weak bees, which, quickly cutting off in the middle with her fangs, she first carries away the rear parts, and soon fetches the other, when

---

123 Virgil

she has bitten off the wings (for earlier carriage) not far from the place where she took it up.

Within a month after her coming abroad, she waxes bold, and ventures into the hives for honey, but because of the strangeness of her voice and appearance, she is spied before she comes near. And at the first (while the weather is warm, and the bees both early and late keep watch and guard at the hive's door) coming singly against many, she is usually repulsed, and sent back again with a flea in her ear, and if by chance she slips in, she does not always escape. Sometimes she is killed in the hive, and brought forth dead, sometimes outside the door, when she has got her prey. But afterwards, the weather waxing cold, and especially in the mornings and evenings, and the bees therefore retiring from the door higher into the hive, the wasps make great spoil, especially among those that are weak. And this they continue until Scorpio, after which time they begin to wear. Nevertheless while they live, that is until Sagittarius, (if abundance of cold and wet rid them not a little earlier) they will be filching, and one wasp will carry out as much as two bees bring in.

The winter wet and cold, kills many of the mother wasps as they lie in their sleep. The spring wet and cold hinders their breeding, for being by that means kept in, when their time is come to fly abroad and feed, they pine and faint, so that either they breed not at all, or very late. And when a warm period in the beginning of Aries having let them abroad, cold and stormy weather comes suddenly upon them, they are shut up again, and so starve for the most part with hunger and cold, that your bees shall not be much troubled with them in such a year. Yes, continuance of wet, though without cold, is such an enemy to the wasps, that in the year 1613 [124], though, the former summer being exceedingly dry, the wasps multiplied, and the winter being mild, the mother wasps were many at first, yet the rain during spring and summer, destroyed their nests, so that there were no small wasps seen until Libra, and then very few.

But the winter when mild, and the spring and summer continuing warm and dry, they live and breed in every place, that, without continual and diligent attendance, you will be sure of a great loss among your bees, though the former year there were but few. For one nest yields breeders enough, if they should all live, to store a whole country.

Therefore, if you love your bees, allow not a wasps' nest about you.

The ready way to rid them is, if they are in a tree above the hole, to smother them with brimstone or puffball [125], as you would kill bees. If in the thatch of a house (when you have made way to the combs) to scald them. If in the ground

---

124 The second specific year reference in FM. See note 76.
125 *Bunt*

(as most commonly they are) you may likewise scald them, and so take the combs out whole, and give the grubs to your chicken, although the boys make better sport in burning them. But if you are in haste, and care only to despatch them quickly and quietly, do this. First, stop their pathways, that those within do not break out upon you (for those that are abroad coming home weary and loaded are more gentle). Then with a wasp spade, search for the nest, which, if it is shallow, is quickly found. When you have found it (which you will know by the easy entrance of the spade) then dig down round about it, and having thus rounded the nest, stamp the earth down on the combs, and so have you done. If you find not the nest because it lies deep, then dig up the ground a foot about the hole, and having found their pathway, stop it fast with earth, and tread in that earth which you dug out, and leave them alone. If this is done in the day when many are abroad, the evening or morning following you may kill them with your foot, but in the evening you may take them all together.

And to destroy those that resort to your hives, place by them cider, soured crab apple juice [126], sour drink or remains, in a short necked open vial or other glass covered with a paper that has a hole in the middle, and so you shall catch many. Also you may take four or five slices or more of sweet apples, or pears, or beast's liver, or other flesh, or anything that they love, and lay them in many several places among your bees on which you shall have sometimes as many as will cover the bait, which you may kill at once, as butchers kill flies.

{*Aristotle teaches another way...*}

The flying moth lies between the hackle and the hive, and breeds little worms, or crawling moths, some in the skirts of the hives, some within and on the stool, wrapped in the dross or scouring of the combs, and some outside on the hive, especially in the cracks of the daubed hives...They also vex the bees with their mealiness, as the snails do with their sliminess. Therefore rid your hives of these guests. The moths are easily crushed in front of, or on the hive, and the snails, though you kill them not, will not long abide, if there is no harbour of long grass, weeds, or other things about the hives. But as for the moth, if you allow her, you yourself shall have more cause than your bees to be vexed. For even though in the cold spring she breeds around the hives, hatching her young by the heat of the bees, yet when the heat of the air will suffice for that purpose, she chooses rather to lay her eggs in woollen, their natural nest and nourishment, especially if it has a raised nap, that there she may safely hide them. In which place, until they are grown to their full size, they lie fretting and eating the cloth, and then after a while they creep out of their skins as flying moths. The maids that sun their clothes to rid them of fleas, let them take heed how they do it near the bee-fold,

---

126 *Verjuice*

lest they bring in worse enemies than they carried out [127]. If the woollen is oily or greasy they like it the better, and for that cause, good housewives' yarn lies not long unwoven.

If ants are near your bees, they will much trouble them, biting them and hanging upon them, although the bees if they are strong, will kill many of them that come to the hive. But if it is a poor stall, they will in time possess the hive, and eat up the honey. The best remedy against them is to scald them.

The spider, like the moth, harbours herself between the hackle and the hive, where usually she has a bee or two in store to feed on, an unsuitable mess for such a mouth. Sometimes she hangs her nets under the stool, which easily entangle a weary bee, when she comes home laden, and missing the alighting place falls into them, yes, and sometimes where the bees are few, chiefly in the winter, they will be bold to enter the hive, and there weave their fatal webs...

The toad is by nature so harmful to the bees, that while he is about the hive, although he lies under the stool, the bees will not prosper. He is said also to devour them at the hive, as the frog at the watering place...

But not any one of these, nor all these together, does half so much harm to the bees, as do the bees themselves...They make the greatest plunder both of bees and honey. For as they of the same hive lie in inviolable peace one with another, so have they no intercourse, no friendship or society with others, but are rather at perpetual defiance, and deadly feud with them. In fighting they are fierce, and in victory merciless; within the space of a day or two, yes, of an afternoon sometimes, if the hive is open so that they have easy passage to and fro, they will have rid it clean. And therefore all bees, of all their enemies, do most dread strange bees, knowing well in what danger they are to be robbed by them both of goods and life.

This robbing is practised all the year. In winter, as often as the weather is fair and warm, some will be prowling abroad. And some are so thievishly disposed, that all the summer long, when an abundance of honey is everywhere to be had for a little labour, they will yet be filching, even though they die for it. In the spring they are more earnest, finding now a suitable time to fetch that which they left behind at harvest, and to repair their decayed store, both of honey and bees. And therefore they now have an eye to them and defend the weaker swarms from their violent irruptions. Those stalls that have lost their queen, or too many of their company, or are vexed with the corruptness of their combs, or dislike their situation because of coldness, moistness, mustiness, bleakness, or unsavouriness, and taking no pleasure in their lives, do now easily allow themselves to be robbed.

---

127 A sentence which provides an intriguing glimpse into sixteenth century domestic life, in which the maids bring out clothes to be rid of fleas, by laying them in the hot sun.

And if none will come to rob them, then on some fair day they will away together, sometimes leaving both honey and young ones behind them.

But in Virgo it is the most dangerous time of all. Then shall all the stalls in your garden be tested of what mettle they are made. And Libra would not be much better, but that the most plundering is done before. Little and poor swarms are now subject to robbing. Likewise those bees that are vexed with the blackness and rottenness of their combs, caused through age, or wet, or blocked with an abundance of harmful decayed honey, will most of them go forth in the swarms, leaving a very few, sometimes not past a handful, in the stock, which yet in robbing-time will keep the door, as though the hive were full, but the robbers finding their weakness, will surely spoil them, if they are not prevented. (How to know such weak bees, and what to do to them, see my notes in chapter 3.)

The robbers are thought to be poor swarms and stocks, which have not sufficiently provided themselves for winter... But indeed such are easier to be robbed, as before was said, than to be robbers. There is no thief to compare to the rich thief, who, though he has enough, and more than enough, yet by hook or crook he will have more, though the poor starve for it.

At the beginning of wheat harvest, the state of flowers decaying, which is always about Virgo, the main robbing begins. Then they send forth some of their stoutest and youngest bees to spy on, and start the attack, which they, going about from hive to hive as far as their walk extends, do start to prove. Where they have once sped, at their return they bring more of their company, until in the end that whole stall is made acquainted with it. Sometimes it happens, that though there are a hundred stalls within a walk, yet the robbery is done altogether by one, sometimes by two or three, all the rest being quiet. And this one thing is strange, that whereas no bees will abide strangers in their hives with them, yet these will allow one another and agree all in stealing, though they are of different hives.

When the thieves, having first made an entry, begin to come thickly, and the true bees perceive themselves to be assaulted by many, they suddenly make an outcry, and issuing out of their holds by troops, immediately prepare themselves to battle. Some keep the gates; some as scout-watches fly about; some run in again to see what is done there; some begin to grapple with the enemy, and that with such a noise and din, as if the drum did sound an alarm. Besides which base sound, you shall soon afterwards, in the heat of the battle, hear a more shrill and sharp note, as it were of a flute, which I am certain is tuned by their general commander, encouraging them to fight for their prince, their lives and their goods. Then shall you see the enemies bestir themselves most adventurously, some violently through the thickest thrusting in at the gates, others scaling the walls, and tearing them down. If they once make a breach, without immediate succour, you shall quickly have an end of that fight. On the other side, the defendants will behave themselves as bravely, not giving any rest to the enemy, partly encountering those that are outside, partly with those that have broken in, whom in a little while they draw out by the heels, some dead and some alive. Likewise outside you may see some slain forthright with the thrust of the spear, some so deadly wounded, that they are not able to go three feet from the place, and some more lightly stricken, and soon to lose the use of their wings, and which for a while leap up and down, forwards and backwards, like mad things…

So loath are these courageous warriors to yield on either side, until there is no remedy…

In their fight they are so furious sometimes, and so thick about in the garden, that, unless you have on your complete harness, you will not dare to come near them.

This also I have noted, that when the robbers are so few, that small resistance will serve; yet being called forth they will not be idle, for you shall see some of them running up and down about the hive, to seek and search if any more come; others, like trained soldiers, practising to fight: here one wrestling with another in single combat; there two or three, or four setting themselves against one, as their usual manner is to deal with the thieves. If you would know whether the fight is in jest or earnest, with fellows or with foes, the manner, and the end of it will show you. For if they are fellows, their fight is not so fierce, and they will part quietly as friends, whereas if they are foes, even though they escape, it shall be with much ado. For if the true men cannot kill the thieves, yet will they hold them by the legs or by the wings so long as they can, in the hope to have help, though they are drawn after. Moreover among the young soldiers, which have scarcely been abroad before, you shall see the elder sort go round about them, smoothing and trimming them in every place, as if they did address, and hearten them to fight.

During the time of this battle, as afterwards, the wasps like vultures prey upon the dead carcasses, carrying them away piecemeal.

The battle being ended by a repulse of the enemy, those corpses, which the wasps have left, they honestly bury as far from the hives as they can bear them… And then they draw together at the city gates, and there they buzz one to another, as if in their language they did talk of the fight, and commend one another for their fortitude.

The robbers, prevailing not that day, will up the next day as soon as it is light, an hour before the bees usually come abroad, and then do they make a fresh assault. The bees finding the enemy among them, are immediately up in arms, and so begins the second skirmish, which, without the taking of the city or the overthrow of the assailants (which seldom happens) continues, until very darkness parts them.

When the true bees, finding themselves overmatched with the multitude, and seeing there is no remedy, and that no resistance will serve, at length they yield and allow the conquerors quietly to plunder their goods. And after a while, when, by being together in the same hive, and sucking the same honey, all smell alike: they will join with their enemies, and help carry away their own goods, and so become friends and live together. At night they lodge with them, but in the daytime they return with their new fellows to fetch that which is left behind. By this means some adventurous stalls are suddenly much increased both in bees and honey, although when a swarm not overstored conquers a poor stall, and so gets, by the victory, more eaters than meat, it becomes their own overthrow, for when their food fails, they die all together.

Seeing therefore that in so cruel and continuing a fight, often the enemies are conquerors, and then all is lost, and if they are vanquished, this victory is not without loss of men and goods, which the enemy ever now and then shifts away. I know your desire is to know how to succour the true men, ether by preventing this dangerous conflict, or by rescuing them in the same. To prevent the conflict, see my nòtes elsewhere. For rescuing them, many practices have been tried. Some {*bee-masters*} cast dust, and some cast drink amongst them, but one does no good and the other harm. For drink makes them all smell alike, so that the true men cannot know the thieves from their fellows, and therefore some do so when two swarms are put together, that they may seem to be of one company. If these usual helps are not helpful, what help is there then? If you perceive their fighting in time before any great harm is done, then this must you do. First close them up, that none can pass either in or out, leaving only a breathing space. Then shall you have a double conflict: one within, and the other outside. The thieves that are within, having no way to escape with their prey, first or last will all be slain. They that are outside, after a little wrestling, seeing nothing to be had but blows, will not long abide this incurable danger. When you perceive the siege to be raised, and that there is little or no fighting outside (which will be about an hour after) then may you let out your bees, making the door half an inch high, and scarcely half an inch wide. Those few that were within will they bring forth to burial: some then, some on the morrow. In the evening, when the bees are all in, shut them up as before. The next day, early, before the bees would be abroad, you must look for some of them again. When they are come, beat them away with a bough, but kill them not, for so may you do your neighbour a shrewd turn, and yourself too.

But let not the bees out before noon, and then make the door so narrow that only one bee may pass at once. So will they keep the robbers out, and follow their business nevertheless. The next day you may let them out rather, and if the door is so narrow that it hinders their passage, you may make it wider. If this does not suffice, but still the strange bees strive to get in, assure yourself, that that stall will yield. And therefore if you are loath to take it now, because of the young bees that may corrupt the honey, then must you look into it carefully, lest by little and little it comes to little or nothing.

But if the bees have yielded before you are aware of it, so that the thieves rob quietly without resistance, or have broken the honey-cells (which you may perceive by the small crumbs of wax upon the stool) then, having shut the hive as soon as you can, the next evening or morning take what is left, otherwise in the end you shall lose all. For the bees which are smelling the honey when the combs are broken, will have it or die for it.

This fierce and cruel robbing being always in harvest, when people are busy in the fields[128], many hives are left honey-less, and they never the wiser. Therefore it is good to leave somebody at home, as well to watch these, as the two legged robbers.

Neither is this robbing hurtful only to those that are robbed, but to the robbers also. For many of them are daily cut off in the assault (you may see them lie sprawling at every hive door) by which their whole stall sometimes is so weakened, that, while they seek to prey upon others, they become a prey themselves.

After a moist spring, when swarms are most plentiful, robbing is most rife; otherwise there is less danger.

Besides those bees which are thus spoiled in robbing, many also are killed by other stalls when they come to them for succour. For in the spring those swarms that were late, or have been half-robbed, when they have spent all their store, the next warm day after, away they fly, some to a tree where they hang until they are dead; some venture into other hives, where if they have a large entrance, they may throng in there suddenly; sometimes they escape with the death of a few, and being mingled together continue with them as if in one stall, but for the most part every one of them dies.

To prevent this loss (1) when you perceive them to wax light and weak, drive them into a stall that has provisions enough; (2) if it be your good fortune to see them entering a stall that is well stored, lift up the hive and let them in together, and so perhaps they may escape; and (3) if you find any hanging abroad, you may put them into what stall you lift, by raising the stall a handful from the stool, and

---

128 Wootton St Lawrence was a small agricultural community in Butler's time, in which the main occupation was in the fields. In 1725, about 290 people lived in the parish; two JPs are listed as residents; there were no schools in the parish. *Parson & Parish in Eighteenth Century Hampshire, ed. W R Ward 1995*

laying the bees upon the table, close to the door.

Lastly, the bees will much destroy one another in swarming time.

Next to bees, the greatest enemies that the bees have is unkind weather, by which at all times of the year both they and their fruits are much impaired.

In summer, extreme heat melts the combs (especially of swarms) and so sheds the honey if the bees are not shaded and well hackled. It also causes the bees to lie out, and so hinders their swarming.

At winter, the sun shining in frosty or snowy weather is dangerous to the bees. For the shine brings them abroad, and the frost chills them, many while flying, so that they cannot return, and many that do return, while they rest on the ground before the hive. But the snow stuns them, and dazzles their eyes, causing them soon to fall, and with its vehement cold to rise no more, and most of all then, when, to shun the wind, they alight in the shade. And therefore, if the snow is lying, the mildness of the weather draws them abroad. It is therefore good to strew the snow with straw, not only in your garden, but also outside the fences, especially in the sheltered side, if conveniently you may, and so shall you save a number, which otherwise you might see lying about like slain men in a battlefield.

Also the freezing easterly winds, and all great frosts kill many in the hives that are open or uncovered, and therefore at such times it is good to shut them up tight, and to see them well hackled.

And if the cold continues to keep them in for a long time, it makes them so sick, heavy and chilly, that many die as soon as they come abroad into the air, unless it is very pleasant.

Also the rain does often soak into the hives and so corrupts the combs, and kills the bees, especially where the company is small, not having enough heat to dry them again. Wherefore ensure that your hackles are always good. And for remedy (if any such chance happen) pull off the hackles on warm days, that the sun may dry the hives again.

But the greatest loss is in the spring. For the bees, especially the young ones (being laden and weary with their labour) some at their work, some on the way home, some at the hive door are beaten down, not only through sudden storms, but also through cold rough winds, and then, unless the sun shines or the wind quietens, they never come home again, so much so that sometimes you may see the lanes strewn with them.

And therefore, when they are afield and they see a storm or rainy cloud arising, immediately they hurry home for life, tumbling to the hive as thick as hail, thronging and throwing down one another before the door for haste. Where, if the cold rain catches them before they can get to the hive, they are in no better situation than those that the storm beats down by the way, although, when they are fresh and light, they will fly abroad in the midst of a warm shower, not caring for it.

They which are thus taken abroad, must take their chance, but if you defend your bee garden as you ought, you shall prevent the fall of many at home. And those that you find chilled with cold (though they are quite dead, without sense, motion and breath, yes and have lain so for all the day) you may, if you are so disposed, revive with the warmth of your hand, so that it will seem a miracle to you. For immediately (their spirit returning) you shall see them begin to pant and breathe again, and soon they will fly away as strong as the best. But if you spy any store of such dead or half-dead bees, then your way is to put them in a glass, and covering it, to turn it round against the fire, until you see them ready to fly.

Also where palm willows, or other trees on which they gather, hang over the water, the rough winds throw down and drown a number of them, while they are at work. Many also, where there are no such trees, when they come but to drink.

For remedy of the first, cut down the trees, and for the other, see my notes elsewhere.

There remains yet another enemy worse than all these. For these all do wrong the bees but by little and little, some in their goods, some in their persons, and there is remedy showed, if industry is not wanting, against them all. But this, when he comes, plays sweepstake with then, carrying away both honey and wax and bees and hive, and all at once, and there is no sufficient remedy found, either in the bee-master, or in the bees themselves against him, neither shall I, with all my skill[129], be ever able to deduce any, unless the wisest of the land, when they meet together, will join with me in the intention. For first the bees are least destitute of their keeper's help, seeing at the times of greatest danger, they cannot be always be under God with him, nor can they conveniently be beneath with him, although some have, for their safety, put this in practice, housing them in and shutting them up tightly all the winter, but not without much inconvenience. For by this means they are debarred of their necessary recreation in a warm house, when it happens, and if by chance they break forth, they lose their way in again, and their lives also. And as they lack in here their keeper's help, so have they no means to save themselves, no, not so much as the silly sheep, which happily may run away. For their resistance, which against some enemies does often prevail, against the violence of this fly Tenebrio [130] avails nothing at all, who stealing upon them while they are at rest, and suddenly surprising them, carries the poor captives (alas) they know not where. Although I have read a 'Story of a Stall' that being stolen did sharply punish the malefactor, making him to submit himself to their master, and to ask him forgiveness. Indeed I will not be bound for the truth of it, for it is

---

129 'with all my skill' – Butler knows his abilities and skills as a bee keeper are well-founded
130 Mealworm beetle

no child of mine, but if any man desire to take it as it, he shall have it as good and cheap as I.

When a thief by night had stolen St Medard's bees [131], the bees in their master's quarrel, leaving their hive, set upon the malefactor, and eagerly pursuing him whichever way he ran, would not cease stinging him, until they had made him, whether he would or no, go back again to their master's house, and there falling prostrate at his feet, humbly to plead mercy for the crime committed. Which being done, as soon as the saint extended to him the hand of benediction, the bees, like obedient servants, did immediately stop persecuting him, and evidently yielded themselves to the ancient possession and custody of their master.

It were to be wished that *pares culpa* [132] might be *pares pane* [133], that all like offenders might have like punishment. But since our bees are not of St Medard's kind, thus to rescue themselves from this mischievous enemy, it is fitting their merit procures them a protection, and as they provide for the health and safety of men, so men should provide for the safety and secure being of them. That such as delight in things for their country so profitable, might not by idle and thievish rogues, unprofitable members of the commonwealth, be discouraged in their honest courses. Wherefore I humbly and heartily entreat all those, whether they are of high or low estate, which shall reap either profit or pleasure by these my pains, that they would endeavour, as much as in them lies, by themselves and by their friends, that against this odious theft it may be enacted, as a law of the Medes and Persians [134] which alters not, that they which feloniously break open these true labourers' houses, shall, like other house-breakers, be deemed and judged as guilty of burglary, and so have no benefit or favour by the Muses, that so violate the Muses' sacred favourites. And here, now my hand is in (though it may seem a hard digression) let me beg a similar advantage for those other necessary creatures, which, for their similar certainty and general profit, the Proverb has joined with them in special commendation to the world.

They serve for man's use both without and within, not only to feed the belly, but also to clothe the back, for which necessary uses, they deserve especially to be loved and defended by all. And yet I think that in any thing, nay, in all other things, there is not so much wrong and spoil done to the country, as in them alone. Sheep stealing, for pity's sake, has now grown so common and so continual. By which means, besides the infinite losses which true subjects daily suffer in that

---

131 Saint Medard (456–545 CE) was a French bishop, whose life was associated with many legendary stories.
132 Equals guilt
133 Like bread
134 Daniel 6.15

kind, the general good commonwealth sustains another great damage in corn, the husbandmen often fearing and forbearing to enfold their land, lest their loss should be greater than their gain. Surely, in my opinion, the very freebooters [135], or highway robbers are more worthy of favour than such. For they are men of more generous spirits, both apt for service themselves, and to breed bold soldiers for their prince and country, who, by good admonition, better implements, or conscience of the sin, are often reclaimed. Whereas these night ravens, for the most part, are base cowardly knaves, neither fit for service, nor labour, a mere burden to the commonwealth[136] and as incorrigible as sheep-biting dogs, which being once fleshed seldom desist, until the bones or somewhat else do happily choke them.

Whoever keeps well both Sheep and Bees, Sleeping or waking, their prosperity will arrive.

---

135 *Boot-halers*, foraging soldiers
136 'commonwealth' appears seven times in the text of FM, where it sometimes means 'the pubic good' (the common weal), but more often it has a deeper religious and political significance i.e. the body politic. The finest body politic is exemplified by the bees – 'Their admirable nature and properties, their generation and colonies, their government, loyalty, art, industry, enemies, wars , magnanimity etc' of the title page of FM.

Chapter 8
Of the Feeding of Bees

Three months of the twelve are rich and plentiful (in which the bees gather abundantly, and store for themselves all the year): Gemini, Cancer, Leo, but Cancer is better than both the others. In Virgo and Libra they live off their daily labour from hand to mouth, little increasing or diminishing their store, unless they fall into the hands of robbers, and then, without rescue, they lose all. But in the other seven, either wholly or partly, they feed upon that store, which the three rich months afforded them. For Scorpio has but the poor gleanings of decaying plants; the three still months nothing at all. Pisces begins to put forth fresh plants, which in Aries and Taurus are well increased, but during this breeding time the mouths are multiplied more than the meat, which unkind weather frequently causes them not to fetch in. So that all this while, they, more or less, spend upon the stock; yes, the weaker stalls somewhat longer, being not so well able to endure the sharp air, and therefore, for want of some store to feed on at intervals at home, I have known some die after mid-Gemini.

For which seven spending months, some swarms are sufficiently provided, which you may reckon as jewels, the very the hope of the flock, whose increase the next year is certain, if they are not over-hived. Some are not provided for half the time, and these, as desperately poor, are not worth the feeding. Others are provided for six, or five, or four months, which by the help of feeding, may live and do well.

Of the first sort are timely, unbroken prime swarms; also fair castings not over hived, before Cancer; yes, and in natural or backward summers before mid Cancer, when the blackberry blossoms have not yet come, nor the honey dews gone, for such have sufficient in both numbers and time, to make sufficient provision.

Of the second sort are the over-little and late swarms, whose paucity and poverty make them too weak to travel for more, and unable to keep what is got or given to them. Such are sure to be driven out before winter is past, by cold and hunger. And therefore if you have omitted to save ones like these by uniting them, yet omit not to save that little which they have and your vain labour and cost in feeding them.

Of the third sort are the middling and indifferent swarms, which by their earnest labour have done well, and gathered a good store of wealth together, but yet, for some want of number or time, the summer leaves them in some need of provision.

As for stocks that have stood two years, and yet have not sufficient stock for these seven months (whether they are such as have not gathered it, or having gathered it have lost it again) they are out of strength, and therefore are suitable to be taken, not fed.

But because unseasonal summers may make good swarms become poor, as contrarily a plentiful summer may greatly improve the worst, after mid Virgo, when bees are to be taken, it is necessary that you test all your swarms, by knocking and weighing them, for the noise will tell you the size of the company, and the weight, their wealth. Those who are used to weighing their hives by hand, will determine the bees' state by guesswork, which until you know to do the same, the scales will direct you. For having taken the true weight of the hive and altogether, if the known weight of the empty spleeted hive being deducted, there remains not fifteen pounds in all, that is to say, for five pounds in bees, the double weight in honey and wax; that swarm is desperately poor, and more suited to be taken than fed. If the swarm with its store weighs between fifteen and twenty, due feeding may save it. If between twenty five and thirty, esteem it as accurate and good, plentifully provided even for a hard winter, and likely to be forward the next spring. And above that the greater the store is, the better increase it promises. Unless, in some extraordinary year, the hive is cloyed with too much, for too much, as well to bees as to men, often does more good than harm.

Moreover, because a long winter and a rough spring following, with some other accidents, may waste those that were good, on the other hand, a short winter and a mild spring may help those that were scantily provided; it shall not be amiss to try again in Pisces or Aries, those that you suspect, and to feed them if you find cause.

The natural food of bees is honey, for want of, or for the sparing of which, many other things have been devised. Aristotle mentions figs, and all such sweet things...And Pliny, speaking more particularly, commends raisins and figs, and teased wool, made wet in sweet wine made of raisins, or new wine boiled, or honey-water...And some of our countrymen have practised giving them basil, bean flower, ground-malt, roasted pears[137], and apples and sweet wort[138]. All of which things, though they will feed on them, yet they cannot be maintained by them without honey.

The manner of feeding bees in their hives is many. Some give them honey in a spoon, but that way many of them besmear their wings, and if their fellows do not lick them clean immediately, before the cold chills them, they die. Others, to avoid this inconvenience, give it to them in a warm toast, but this way wastes the honey, and does not altogether avoid the former inconvenience. Others have other devices. But indeed the only good way to feed bees is with a comb, after

137 Warden - an old variety of baking pear
138 An infusion of malt or other grain which after fermentation becomes beer

this manner. First, take a fresh comb of live honey out of a hive, and lay it on some prop or stay, that the bees may work, as well as under or upon it. This prop may be a wooden grate seven or eight inches square, made of two sides half an inch deep, and three ribs fastened into them with dovetails, or with small nails. Where there is a need two several square sticks may serve, but then you cannot so quickly either set it in, or take it out.

Then in a fair calm evening (when the heat of the day, and the bees' work is past) place this on the stool, so that the feeding comb is reared as near the hive combs as may be, not touching them, lest the bees fasten this and them together. Then see that the hive, set down in its place, is closed everywhere, and at the door only leaving room enough for a bee or two to pass. Then will these bees work afresh, not ceasing day or night until they have rid the comb clean, which within forty eight hours will be effected. If they need anymore, the next evening do likewise. But always, when it waxes dark, and the bees are all in, bar up the door, and do not let them out until the next evening, when other bees are quiet. Or if you do it in the morning, be sure also to take out the comb, whether it is cleaned or not. And still leave the hive closed, with a narrow passage.

If your poor bees should not in this way be defended from strangers, the feeding of them would prove a starving of them. For the borderers smelling the booty will be sure to have part of it, if they can come at it, and when that is done, they will set upon the other, and so spoil all, as often it falls out through this carelessness. Which causes some to condemn all feeding of bees, as painful and fruitless, saying, 'If you do not feed them, they can but die, and so will they do when you have bestowed your labour and cost'. But this is disproved by experience, for those, which being suitable to be fed are thus fed, do seldom miscarry. That summer they provide sufficiently for winter, and the next they are as likely to swarm and be fat as another.

You may also feed your poor swarms together (if you have no neighbour- bees to beguile them) by setting any refuse honey or leavings abroad in your garden, having first barred up those that do not need it. This feeding honey, like that which is stolen, when they have first taken their refreshment, they convey into their empty cells, which, because they cannot now shut them up, as before Virgo, for want of wax, they do but half fill. And therefore they first spend this lately gotten honey, refusing that which was more safely laid up, unto the last.

It is good to feed bees before they need (that they may save their store, which they have shut up in their cells, until the spring) particularly in the later part of Virgo, when the combs are taken, or in Libra. For those that have spent their own store, and have little or nothing left in the end of winter, are so discouraged and so feeble with fasting, that knowing their thin bodies can bear out no cold, they will not come abroad, but when they are fed, unless the weather is exceedingly warm and calm, and the more they keep in, the weaker still they are, and less apt to breed. But those whose early feeding has caused them to spare their store until the spring, will be as cheerful as the best; in any reasonable weather they will abroad, and fetch in that fruitful Ambrosia, which causes them soon to increase and multiply.

At this first time therefore first finish the public feeding, and then begin the private, especially of those that are under eighteen pounds, to which if you give now the better part of their due allowance, you may provide for their deficiency, as also of the rest, at the second feeding-time, when their need will better appear.

In Pisces or Aries, as soon as the weather is warm (not before, lest the cold chill them in their work) if you fear they will be lacking (which you may perceive by their lightness and unwillingness to come abroad) supply their need again, and again, if need be. But in this second feeding, for lack of a honeycomb, take a dry comb, reserved for the purpose, and pour onto it as much honey as it will receive. If you think it is not liquid enough, then either warm it first over the fire, or else spread it all over the comb with your finger, that it may link into the cells (for which purpose live honey is best) then use this honied comb as the honeycomb.

If either these fed bees, or any other, chance afterwards to need feeding (especially in Taurus, or a little before or after) then feed them daily until mid-Gemini, giving them every evening or early morning, a spoonful of honey, and taking away the comb again before other bees are at work. But this is to be done without intermission, for the bees will duly look for it, and languish, if once or twice they are not given it.

By this means I have saved swarms that forsook their hives for hunger, hiving them again in their own hives, which proved good the same year.

Chapter 9
Of the removing of Bees

In the removing of bees be careful to avoid these five evils: (1) hindering their swarming; (2) their honey gathering; (3) breaking their combs; (4) robbing; (5) loss of bees.

Remove always on a fair day and, as near as you can guess, in settled weather. For when they are moved to another place, if it is within their circuit or walk, they will fly to their old standing place as soon as they are let go and linger about it three or four days, and sometimes longer, where, if the cold catches them, many lose their lives. And if you remove them out of their familiar surroundings, then, confused in an unknown place, they fly about for a while viewing the country, and searching for their old home. When they are weary, they rest anywhere, and if foul weather comes upon them, they are in similar danger.

For the time of the year, remove not in the three still months, or in a fortnight before or after, for fear of losing the bees. For if foul weather falls not, the very still cold will kill many, while they are straying abroad, and of those that return, not yet being acquainted with the hive door, some will fall short, and some upon the hive, where, while they rest panting, the cold chills them.

Taurus, Gemini and especially Cancer will frustrate and hinder the swarming, as well as their honey gathering, and Cancer brings the danger also of breaking their soft combs.

In Leo, though the swarming is past, and robbing time not yet come, yet there remains some honey gathering and the combs being then most weighty and most weak, the danger of breaking them is greatest.

To remove in Virgo (when the bees are vying to be masters) is dangerous for robbing. For the native[139] or old inhabitants of the garden (as they go about prying for booties) finding new neighbours come among them, will be sure to visit them, and while the chief of their strength is straggling abroad, seeking their old dwelling, they will bring the rest such cheer to their housewarming, as shall happily make the house too hot for them. And then they will be pleased to go along with them, and help to carry their own goods after them.

The most suitable time is either in Libra, and the forepart of Scorpio, that they may thoroughly know their new standing before the weather is too cold, or in Aries, and the later part of Pisces, that they may be acquainted with it before much gathering of honey.

---

139 *Indigene*

Nevertheless, if you have bees in other men's keeping, whose care and skill you mistrust, you were better to remove them unseasonably with some loss, than to hazard all by their ignorance and negligence[140].

But if you may choose, remove in Libra only, which is simply the best.

And for the removing of a swarm into another garden (whether it is near or far off) the only time is the evening or night next after the hiving, that it may be at its new standing, ready to work, in the morning, and so lose no time, nor break its first comb in the carriage.

In the evening, when you mean to remove, an hour before sunset prop up the hive from the stool, with three bolsters two or three inches thick, that the bees may ascend from the stool. About half an hour after, having prepared another stool of the same height, and covered it with your mantle so that the middle of the mantle is over the middle of the stool, move the stall with its stool, if you may, a little aside, and set this cornered stool in its place, or, if it cannot easily be moved, then set the cornered stool close to the old stool, either beside it, or before it. This done, lift up the stall from its old floor and set it upon the new, and then wiping the bees from the old stool (if any remain) with your brush, either take the stool away, or cover it with a cloth. And then if your new stool is only a plank without legs, borne up by some other means, it is good to set it upon the old. Within a while when the bees are all in, knit the mantle at the four corners over the top of the hive, so that the knots may not slip, and immediately bind it to the hive about the middle slackly with a small line, and wrest it fast with a little stick. And so is the stall ready to be removed.

They are used to commonly make no more ado, but after sunset when the bees are at rest, lift up the stall, and set it upon a mantle spread on the ground, and so bind it up, leaving the bees on the stool (which in a good stall are not a few) behind them. Either way, for those stalls which have all their bees up in the hive, may serve well enough.

The best way to carry your stall is on a cow-staff between two.

If it is light, one may carry it in his hand. But however, be sure it hangs perpendicularly for fear of breaking the combs, especially if you chance to remove before Libra, when the wax is soft, and the lower parts of the combs are heavy with young bees, as well as the upper with honey.

When you have brought the stall home, you may let it stand bound as it is, all night in the house. On the morrow, when the weather serves, set it on its seat, but if it is foul all the next day, keep it still bound until it is fair. And then having loosed the line, and taken away the mantle, cloom it up immediately, leaving it for three

---

140 Butler is clearly dubious about the competence and practices of contemporary beekeepers, which is why he has written FM.

or four fair days, with a very narrow entrance, for fear of robbing. For their new neighbours, even now also (though not so eagerly as in Virgo) will test them, and they will not so stoutly resist, until they are acquainted with the place.

Chapter 10
Of the fruit and profit of Bees

In which is shown first the gathering[141] or taking of combs; secondly, the testing of the wax and honey, with the making of meth[142] or hydromel[143], and thirdly, the unique virtues of them, for the use and comfort of man.

The first part of this chapter shows the taking of the combs.

The most usual, and generally most useful manner of taking the combs, is by killing the bees. For which the natural and seasonable time is the latter part of Virgo, from the end of the dog-days until Libra, because until then the combs are full of young bees, which deceive the honey-men, making the hive heavier and the honey worse (for the young bees as well as solid cell debris[144] spoil the same...) and after that time, the weather waxes colder, not so suitable for the running and working of the hive, and the honey is likely to decrease, either by their own spending or by the spoiling of robbers. Except in the heath countries, where their gathering lasts longer, for there they defer their taking until mid-Libra.

At this time, therefore consider with yourself what stalls you will kill. Swarms that may live, yearlings and two yearlings that are yet to prove themselves, keep for store. Similarly, those that get rid of their drones early, and especially those that draw out their young drones. Those of three or four years, which, by reason of their not swarming this last summer are full of bees, likely are fat, and therefore worth taking, but they are also good for store, unless the frequent honey-dews have made them over-fat. But those of that age which have cast twice (except they were very forward and had beaten away their drones early) are not likely to continue, and therefore are to be taken.

Likewise all poor swarms unworthy to be fed, and all light stocks whose stocks are decayed. For they will surely die. Neither is it safe to trust any after they have stood five years, unless it is some special kind of bee, which casts often, and yet beating away their drones early, do still keep themselves in heart. For such I have kept nine or ten years, and I have heard of some of a greater age. Moreover, all stalls of three years old and upwards, that have missed swarming two years together, and especially those that having lain forth the summer before, did not cast this last summer, for such do seldom prosper afterwards. It is therefore better to take them now while they are good, than in a vain hope of increase, to

---

141 *Vindemiation*, the gathering of grapes or other fruits
142 *Meth.* abbreviation of *Metheglin*, a spiced or medicated variety of mead, originally esp. popular in Wales.
143 *Hydromel*, a mixture of honey and water, which when fermented is called vinous hydromel or mead.
144 *Coome* – blackened and/or old honey which stops up the cells

keep them until they perish. Likewise if you have any that are very fat and full of honey (as some years some will be, even down to the stool) those are ripe and ready to yield their fruit. One such is worth three or four. Take them therefore in their season, for wanting room to breed in (their cells being full of honey) they will decay by little and little, and consume to nothing. And therefore as in a wet hungry year you must keep the best, so in a dry year, rich and plentiful in honey-dews, the worst are likely to prove best for store.

But generally, take the best, and the worst. And always suspect those that do not get rid of their drones in time.

Also those which the robbers do eagerly assault, and if their combs be once broken, delay not their taking.

Having made choice of your stall to be taken, some two or three hours before sunset, dig a hole in the ground, as near the stool as may be, about eight or nine inches deep, and almost as wide as the hive skirts, laying the small earth round about the brims. Then having a little stick slit in one end, and shaven at the other, take a sulphur[145] match, five or six inches long, and about the size of your little finger, and making it fast in the slit, stick the stick in the middle or side of the hole, so that the top of the match may stand even with the brim of the pit, and then set another by it dressed in the same manner, if that is not sufficient. When you have lit these matches at the upper ends, set over the hive, and immediately shut it tightly at the skirts, so that none of the smoke may come forth. So shall you have the bees dead and down in less than quarter of an hour.

Next to sulphur is the smoke of bunt or great puffball[146], touchwood[147], or mushrooms, used in a similar manner, but they are neither so quick, nor so sweet. And when required, some smother them with dank straw or hay, but then the honey will smell of the smoke. And therefore some drown them in a tub of water, but that hurts the honey, and does the hive no good, and besides that, many of the bees not being quite dead, will sting those who handle the honey.

The bees being dead, carry the honey into the house. If any bees escape, they will die that night, but if you fear they will do any harm, you may kill them immediately on the stool.

Another way to take the combs is by driving the bees. The manner of it is like this. At midsummer, or within two or three days after, in a fair morning an hour before sunrise, lift the stall from the stool, and set it upright and fast on the ground in a brake[148] with the bottom upward, and quickly cover it with an empty

---

145 *Brimstone*

146 *Puckfist*

147 *Touchwood*, the soft white substance into which wood is converted by the action of certain fungi, and which will burn slowly

148 *Brake*, supports set in the ground to receive the skep

hive, having first laid two spleets[149] on the bottom of the full hive, so that the empty hive may stand the faster. Then wrapping a mantle round about the chinks or meeting of both the hives, and binding it fast with a small cord above and beneath, that a bee may not get out, clap the full hive or remover round about a good many times, pausing now and then a little between, so that the bees may ascend into the empty hive. And when you think that most of them are driven up (which will be about half an hour after) set the upper hive or receiver upon the old stool.

Provided always, before you go about this business, that all the stalls in your garden are first closed, lest they trouble you and your poor bees.

This kind of taking is much applauded at first, because men think thereby to save both bees and honey, but it turns out with them as it is in the proverb, 'All covet, all lost'. For the honey is neither so good, as being not yet in season, and it is also spoiled by the young bees, which cannot easily be cleanly removed from it, neither so much by almost the one half, since there remain yet six or seven weeks of honey gathering.

And the bees, as men forcibly driven from their goods and children, are so discouraged, that they seldom thrive after it, especially those that have swarmed, seeing their company is now left small, and the after-brood is destroyed, which should have supplied the rooms of those that are gone. And as for those that have not cast, they might after that time yield a swarm, which would be better than the whole stall being driven, and if they did not swarm at all, they would be so much the better, either to take for honey, or keep for store.

This driving of bees into empty hives being not so profitable as it seems, I would rather commend to you the driving of one stall into another, by which the fruit of one is taken, and the lives of both are saved together.

And thus some are to be driven in the latter part of Virgo, when they have finished breeding, and some in Aquarius or Pisces, before they begin to breed again.

In Virgo such stalls only are to be driven, as are fit to be killed, and that into yearlings or two yearlings, which that year have cast twice, and therefore have few bees left in them, but yet have honey enough. The manner of it is this. Having first placed these two stalls, the Remover that is driven and the Receiver, as near as may be to one another, and so let them stand together six or seven days, until they are well acquainted with their standings, when you see the weather fair and constant, late in an evening, about ten o'clock, set the Remover fast on the ground in a brake, with its bottom upward, and the Receiver upon it, and bind

---

149 *Spleets* – small split sticks inserted into hive, to stop the walls collapsing under the weight of honey. See R J Hawker, 'The Earliest Record of Beekeeping in Northern England' (2015) p 14

them close together, as in the former driving. And then, by often clapping the Remover between your hands about the space of a quarter of an hour (now and then pausing between) having driven most of the bees into the Receiver, and so mingling them all together, let them so stand until the morning. In the morning, about sunrise, if the weather is fair (otherwise you must stay longer) do the same, having first shut and covered the other stalls.

This done, set the Receiver on the Remover's stool, but be sure to bolster it up with three tile-shards, so that the driven bees may easily get into the hive on every side. And then knock the remover down onto a table two or three feet square, set close to the front of the stool, and, by clapping of the hive, immediately get as many of the bees out as you can. And forthwith carry the Remover about five and a half yards[150] from the stool, and there lay it down, so that the combs lie edge-long; after a little while clap it twice or thrice, which will make many of the bees fly forth. Then remove it to another place about the same distance, and there do the same, and so to another, and another, until few or no bees come forth by this means. And whenever you come to a new place, and there have got out some bees, leave there the Remover, and go directly to the Receiver, and a little beyond, for the bees will follow you, and thereby the sooner recover the hive.

After this, having removed the Receiver again, and laid it with the combs edge-long, as before, stay until you see the bees ascended to the highest part of the combs in the skirt of the hive, and then resting it on the edge of a tub[151], and turning the bees towards your readiest hand, with two or three claps force them out into the tub, and then immediately carry the hive to another place, and when you see more bees ascended, have it back again to the tub, and there clap them out as before. This repeat as often as you see any store arise to the upmost part of the hive skirt. Which, when they cease to do, the hive is very nearly rid of its bees. At intervals, carry the tub to the stall, and knock out the bees onto the table, then, having first loosened the[152] spleets' ends, take out the combs, beginning at one side, and even when you have taken out a comb, wipe off the bees with a feather of a goose wing into the tub, and send it in, out of their sight. When the combs are all gone, set the hive and tub before the Receiver, that the bees may take up your leavings. As soon as they begin to be quiet, take away the bolsters, and cloom[153] up the hive very tightly, leaving the door no wider than must needs be. And when all is done, set open your other stalls, and carry the hive and tub from among the bees.

---

150 *One perch*
151 *Kiver*
152 See note 150
153 See Glossary

If you think there is not sufficient provision for this double stall in that single hive, place a full comb or two as need requires, of the Removers upon them, and so will your bees delight and prosper in new wax, which in old spoiled combs would decay.

In Aquarius or Pisces, when you have weighed your hives, those that you find by their lightness unlikely to endure the spring for lack of food, you may in similar manner drive into such provided stalls, as have fewest bees, and so will those Receivers be much the better, and cast both earlier and larger swarms. And if by chance, at any time after, you find a stall decayed, thus may you save them. Otherwise, if it is suitable to be fed, feed it.

If, the weather not being warm, you find some bees chilled about the hive, fill your warm hand full of them, and soon they will fly away to their fellows. And if any happen by chance to sting you (which they will seldom do) your hand will have more ability to revive the rest.

This driving will not be so troublesome as the former, because the poor bees will easily change their hungry home for a place of plenty.

Cutting[154] the combs or taking away a portion of honey[155] is a third kind of taking, which is the cutting out of part of the combs, part being left for the bees' provision. And this was to be done at two times in the year {*according to Columella...*}

But what part is to be taken, and what left, I find it not determined... This way of taking, it appears, was anciently used in plentiful countries, such as Greece, Sicily, Italy etc. But the former method, namely, in the spring, Aristotle nowhere mentions, and surely it must do more harm than good, seeing the hives are then full of young bees, which being slain, puts an end to their swarming, and the store of honey, which they seek, is then well spent.

And that also in the autumn (which yet is the better time) seems no less unprofitable than troublesome, because the bees in the spring following, even if they do not lack honey to live on, yet will they lack cells to lay their young in, by which their breeding will be hindered. And at neither time can it be done without considerable spoiling of the bees.

But however it fared with them, for our country I take it to be very unsuited. And therefore I say the less of it, referring the curious reader to the fifteenth chapter of the ninth book of Columella, and to Georgius Pictorius[156], who in his fourteenth chapter writes of this at length.

---

154 *Exsection*
155 *Castration*
156 Georg Pictorius of Villingen, Germany, physician and author, c. 1500-1569

The second part of this chapter shows the trying of (1) honey and (3) wax, with the (2) making of meth or hydromel.

The hive being housed, squat it gently against the ground, on the sides, not the edges of the combs, and loosening the ends of the spleets with your fingers, and the edges of the combs, where they stick to the sides of the hives and with a wooden slice, take them out one after another. Then having wiped off the half-dead bees with the feather of a goose wing, break the combs immediately, while they are warm, into three parts – the first, complete honey and wax; the second, honey and wax with pollen[157]; the third, dry wax without honey. And that they may break right where you would have them, mark the places deeply with the edge of your knife. But first provide necessary instruments such as pans, tubs, tongs, wide sieves, or wheat-ridders, a slice, knives, straining bags, a tub or keeve[158], with a tap, and tap-waze[159], a hair fine strainer[160], honey pots, wax moulds, meth barrels etc.

These things provided, take out the first comb, and setting the honey end in a sieve[161], resting on tongs over a clean pan or wooden tub[162] that will not leak, mark and break off the first part for honey, and leave it there, then going to the sieve fitted with a tap and tap-waze, mark and break off the second part for meth or hydromel, and leave it there, and lay the third part aside for wax. Then taking out another comb do the same, until the sieve is full.

If you intend to make two spurts, and so two sorts of honey, let your assistant without delay cut the first part into thin slices, and, without any more ado, let the honey run into the first spurt. But this is to be understood of the darker part of the comb, for the pure white cells in the upper part (which contain nothing but pure white, or yellowish live honey) you may as well crush between your hands, and this will be fine ordinary honey.

But if, for some special use, you would have some honey yet more fine and pure, then only slice the purer part of the combs, being yet warm with the temperate heat of the bees, and so let the pure live honey run through a clean sieve…all honey which runs itself (as new wine and oil) is called *Acceton*, is most commendable.

This *Acceton* or finest nectar, for its unspoiled purity, is called virgin honey…

Of which there are two sorts. The true virgin honey is of a swarm; that which is of an old stall, though it runs first and of itself, and was gathered the same year, yet being partly mixed with other honey, and laid up in decayed vessels, not in

157 *Sandarach*
158 Vat for holding liquids
159 A strainer placed over the tap-hole in a mash-tub, to prevent any solid matter from passing into the tap.
160 *Hairen clensieve*
161 *Ridder*
162 *Kiver*

the pure virgin cells, is but a second or bastard virgin honey, rather better to be called the finest ordinary.

But the hard corn honey in the top of the combs, especially if there is any quantity, because it will not run, you must either wash into the warm meth liquor, or melt it with the cells on a low fire, or in a hot oven…and so shall you have the honey by itself, and the wax swimming above it, which you may take away when it is cold. But this good honey will become coarse, and therefore put it to the second spurt.

Having now taken so many stalls as you can dress this evening, take the rest as soon after as you may, and let the honey be all tried out, before you soak the second part.

The hives being emptied, carry them into your garden (a perch at least from any stall) for the bees to take up your leavings, and have still by you a pail of fair water to wash your hands in, which water must be for the meth.

When the honey has run what it will, put this first spurt, whether it is ordinary or virgin honey, into a picked bag, to strain it into its pot by itself. And crush the remainder with warm hands so that it may run again for a second sort, which is similarly to be strained. That which is left at the last, in the bags, sieves and elsewhere, wash into the second shoot or spurt of the must[163], to give it its true strength.

The weather being not warm, set the honey by the fire to help the running.

Otherwise, if you are in haste, and mean to make only one sort of honey, first slice off the upper part of the comb (even as much as you find empty of pollen) for honey, and immediately let your assistant work it all together with warm hands, and so make but one spurt, which afterwards is to be strained. Then going to the sieve, slice off the second part (even all that has honey) for meth. And set aside the dry part for wax. And thus will your honey be good enough, and so, compared with the common honey, may well be considered fine.

For the honey men (because to cut each comb into several parts, and in different ways to dress each part, would be too tedious for those that have much to do) make but one work of all, with a thin light shovel pounding and compounding the honey, and wax, and bees, and young bees and pollen all together. And then putting this mixed stuff into a strong hair bag, with a press or strainer[164] violently wring out all that will run. And this, having first its season of heat over the fire, they put up into barrels or other vessels to work, when though it is greatly purged, yet can it not choose but to participate in the nature and taste of those things with which it was so thoroughly infected. This done, the mixture remaining in the bag

---

163 The thick, pulpy mixture prepared for fermentation
164 *Wrenge*

they slice with a shredding knife into a trough or other vessel, and wash it and mash it in fair water for mead, which, when the sweetness is all washed out, being crushed dry, the balls they strain for wax.

Honey when put warm into pots, will in two or three days work up a scum of wax, honey and dross together, which being shaken off with a spoon, can be added to the second part. In cold weather the honey will not work well without the heat of the fire. The best way is to put it into an oven after the batch is forth, but not before you can bear to hold your hand on the bottom, for fear of over-heating the honey. The next way is to stir it in Balneo Maria[165] until it is all warm.

The differences and degrees of honey in goodness are natural as well as artificial. For as it is made better or worse by the ordering and handling of it, so is it in itself better or worse according to the different condition of the soil where it is gathered. The champian[166] honey is accounted almost twice as good as the heath honey, although they are both prepared similarly. For when the common champian is sold for nine pounds a barrel, the similar heath honey will scarcely yield five. And generally the finer the wheat and the wool is, the finer is the honey of the same region, and therefore it is no wonder that the coarse heath honey is as much like coarse honey as it is wool.

Good honey, when it has wrought, has these properties by which it is known: it is clear, sweet-smelling, yellow like pale gold (but true virgin honey is more crystalline at the first) sharp, sweet, and pleasant to the taste, of a moderate consistency between thick and thin, so clammy that being taken up on your finger's end, in falling it will not part, but hang together like a long string, as that honey does which has been clarified…

This good honey, especially that part which is in the bottom, will in time grow (like corn honey, in the uppermost part of the combs) hard and white, like the honey of Spain and Narbona in France, which is accounted the best, and is compared with that of Hymettus and Hybla. But this is to be understood of ordinary honey, for the pure virgin honey will be neither hard nor white, but changes its liquidity and crystalline clearness into a thick softness, and bright yellow colour.

And always the best part of all honey is that which is lowest in the vessel. For as the best oil is in the top, and the best wine in the middle, so the best honey is in the bottom…

The weight of these three, one to another, has this proportion. Oil is not so heavy as wine by one tenth part, for if you fill a measure with wine, and divide it into ten parts, the same measure of oil is no heavier than nine of them. And

---

165 Later termed *bain-marie*, lit. 'the bath of Mary,' a warm water bath so called from the gentleness of this
      method of heating.
166 *Champain* - from open, level country

honey is heavier than wine by the half, for if you fill a measure with wine, the same measure of honey will weigh that and half so much more…

The second part of the combs, appointed for hydromel or meth, you must first rid of the pollen as much as you can, cutting off that which is by itself, and picking out that which is among the honey, all of which refuse, because of the wax that is with it, is cast to the third part.

In the morning let this first spurt of the must (or wort), being made of its true strength run through the tap-waze. The pulse which remains, when you have squeezed out the liquid, break and wash in fresh warm water in the sieve, for a second spurt. When it has lain awhile, soaking, first take those parcels that swim, and squeezing out the liquid between your hands, lay the balls aside to the third part (but let your bees go through them) then take up those that lie in the bottom, and do likewise, which because they have most honey, you must take the greatest care in washing and crushing them. And while this is doing, let this small liquid run into a vessel by itself. When it is out, wash into it all the remainders of honey, adding some coarse honey, if need be, to make it of its full strength, and then let both streams run together through a clean sieve into the vat again. And thus you shall lose none of your honey.

Meth or hydromel is of two sorts: the weaker and the stronger, mead and metheglin.

For the making of mead, if the must, when it is all together, is not strong enough to bear an egg the breadth of a two pence above it, then put so much of your coarse honey into it as will give it that strength, which is sufficient for ordinary mead. And afterwards, until night, every now and then stir it well above the vat.

If you would make a greater quantity, then must you add a proportionate measure of water and honey, namely six of that for one of this. The learned physician Mathias de Lobel[167] requires this proportion of six to one to be boiled to four. His recipe of spices is cinnamon, ginger, pepper, grain, cloves. The second morning put to the must the scum of the honey, stir all together, and stoop the vat a little backward. When it has settled an hour or two, draw it out to be boiled. And when you see the grounds begin to come, stay, and let the rest (except the very thick grounds, which you should cast to your bees) run into some vessel by itself, which, when it is settled, pour out into the boiling vessel through the fine strainer and cast out these grounds also into your garden.

This must being set over a gentle fire, when you see the scum gathered thick all over, and the bubbles at the side begin to break it, having slacked the fire, to cease the boiling, skim it clean. Then without delay make a fresh fire to it, and

---

167 Flemish physician and botanist, 1538 – 1616, and for a time, James I's physician

when you see the second scum ready, having slacked the fire again, take it quickly away, then make to it the third fire, and let it boil to the wasting of a fourth part, if it is made of the washing of combs, and to the wasting of one fifth or sixth part, if it is made of clean honey, not ceasing in the mean space to take off the scum as clean as you can. One hour's boiling may suffice, but if the meth is of clean honey, it may as well be done in half the time.

After all this, put in the spices, viz. to a dozen gallons of the skimmed must, ginger one ounce, cinnamon half an ounce, cloves and mace each two drams[168], pepper and grains, one dram, all well-ground, the one half of each being deposited in a bag, the other loose, and so let it be boiled a quarter of an hour more.

The purpose of boiling is thoroughly to incorporate the water[169] and the honey, and to purge out the dross, which once being done, any longer boiling is unprofitable, diminishing more the quantity, rather than increasing the strength and goodness of the hydromel.

As soon as it has boiled enough, take it from the fire, and let it cool; the next day, when it is settled, pour it out, through a hair sieve or linen bag, into the vat (reserving still the sediment for the bees) and there let it stand covered three or four days until it works, and let it work two days. Then draw it through the tap-waze, and turn it into a barrel scalded with bay leaves, making the spice bag fast at the tap. If there remain many grounds, you may purify them by bottling and skimming again as before, but this will never be so good as the first, and therefore you may put it by itself, or with some remainder of the best, into a small vessel to spend first, before it becomes sour. If the meth is not much, you may put it into a cask the next day, and let it work in the barrel. Being casked, it will in time be covered with scum[170], which if, by jogging the vessel, or by other means, it is broken, the meth will turn sour. But if so, it will make excellent vinegar, and the sooner, if it is set in the sun; the longer you keep it, the better it will be.

Metheglin is the more generous or stronger hydromel, being similar to mead as wine is to lora[171]. For it will bear the weight of an egg, the breadth of a groat[172] or sixpence, and is usually made of finer honey, with a lesser proportion of water, namely, four measures for one, receiving also in the composition as well certain sweet and wholesome herbs, and also a larger quantity of spices, namely, to every half barrel or sixteen gallons of the skimmed must, eglantine, marjoram, rosemary, thyme, winter-favourite, half an ounce; and ginger two ounces; cinnamon one ounce; cloves and mace half an ounce; pepper, grains, two drams; the one half of each being bagged, the other boiled loose. So that whereas the ordinary mead

---

168 a dram is one eighth of an ounce
169 Boorne i.e. water from a stream or brook, used in mead making
170 *Mother*
171 Pomace wine, a weak fruit wine
172 *Groat* - an English silver coin worth four old pence, issued between 1351 and 1662.

will scarcely last half a year, good metheglin the longer it is kept, the more delicate and wholesome it will be, and afterwards the clearer and brighter, according to the earlier meaning of the name.

He that wishes to know the many and sundry makings of this wholesome drink, must learn it from the ancient Britons, who in that respect surpass all other people. One excellent recipe I will here recite, and it is of that which our renowned Queen of happy memory did so much like, that she would every year have a vessel of it[173].

First, gather a bushel of sweet-briar leaves, and a bushel of thyme, half a bushel of rosemary, and a peck of bay leaves. See that all these are well washed in a furnace of fair water. Let them boil for the space of half an hour, or longer, and then pour out all the water and herbs into a vat and let it stand still until it is but milk warm, then strain the water from the herbs, and take to every six gallons of water one gallon of the finest honey, and put it into the spring water[174] and labour it together for half an hour, then let it stand two days, stirring it well twice or thrice each day. Then take the liquor and boil it anew, and when it does, skim it as long as there remains any dross. When it is clear put it into the vat as before, and there let it cool. You must then have in readiness a vat of new ale or beer, which as soon as you have emptied, turn it upside down, and set it up again, and immediately put in the metheglin, and let it stand three days, fermenting. And then turn it up in barrels, tying at every tap-hole, by a pack thread, a little bag of cloves and nutmeg[175] to the value of an ounce. It must stand half a year before it can be drunk.

The third part consisting of wax and dross, set it over the fire in a kettle or cauldron that may easily contain it, and pour into it as much water as will make the wax to swim, that it may boil without burning, and for this cause, while it is seething with a gentle fire, stir it often. When it has stood a while and is thoroughly melted, take it off the fire, and quickly pour it out of the kettle into a strainer of thin strong linen, or of twisted hair, ready placed upon a sieve or press, and then winding and doubling the neck of the bag, lay on the cover and press out the liquid as long as any wax comes into a tub of cold water, but first wet with that both the bag and the press, to keep the wax from sticking. At the first comes forth most of the water; at the last it is mostly dross; in the middle, it is mostly wax.

The wax becoming hard, make into balls, squeezing out the water with your hands. When you have this done, soon while they are warm break all the balls into small crumbs in a small cooking pot[176] or kettle set over a soft fire. While it

173 Elizabeth 1 died in 1603
174 *Boorne*
175 *Mace*
176 *Skillet*

is melting, stir it and skim it with a wetted spoon in cold water, and as soon as it is melted and skimmed clean, take it off. And having provided the mould, first warm the bottom, especially if the cake is small, and smear the sides with honey, and then instantly pour in the wax (being as cool as it may run) through a linen straining bag. When you come near the bottom, pour it gently until you see the dross coming, and strain that into some other mould by itself. And when it is cold, either try again, or having pared away the bottom, reserve it, as it is, for some use.

When the wax is in the mould, if any froth yet remains on it, blow it together at one side, and skim it off lightly with a wet spoon.

This done, do not set the cake elsewhere, where it may cool hastily, but in the warm house, and if it is large, cover the mould with a dish, as closely as you can, to keep the top from cooling, until the inward heat is allayed, and so let it stand, not moving the mould until the cake is cold. If it sticks, a little warming of the vessel or mould will soon loosen it, so that it will slip out.

The properties or tokens of good wax are (1) mostly yellow, sweet, full-bodied; (2) firm and of good texture; (3) light; (4) pure, and void of all other matter…

The third part of this chapter shows the singular virtues of (1) honey, (2) meth, and (3) wax for the use and comfort of man.

Honey is (1) hot and dry in the second degree; it is of (2) subtle parts, and therefore pierces like oil and (3) easily passes into the parts of the body; it has (4) a power to clean, and in addition some sharpness, and therefore it (5) opens obstructions; it (6) clears the breast and lungs of those humours[177], which fall from the head to those parts; it (7) loosens the belly; (8) purges the foulness of the body, and (9) provokes urine; it (10) cuts and casts up phlegmatic matter, and therefore sharpens the stomachs of those who, because of this, have little appetite; (11) it purges those things which hurt the clearness of the eyes; (12) it nourishes very much; (13) it breeds good blood; (14) it stirs up and prefers natural heat, and prolongs old age; (15) it keeps all things un-decayed, which are put into it, and therefore (16) physicians temper with it such medicines as they mean to keep long; (17) yes, the bodies of the dead, being embalmed with honey have been thereby preserved from putrefaction; (18) it is a sovereign medication both for outward and inward maladies; (19) it helps injury of the jaws, (20) the embrasures growing in the mouth, (21) and the festering or inflammation of the muscle of the inner gullet, for which purpose it is gargled, and the mouth washed in addition; (22) it is drunk against the biting of a serpent, (23) or mad dog, and (24) it is good for those who have eaten mushrooms, (25) or drunk poppy[178], against which evil, rose-honey is taken warm. (26) It is also good for the

---

177 Any of the four fluids of the body which were believed to determine the state of health and the temperament of a person or animal.

178 *'Drunk poppy'* – presumably milk of the poppy used as an opiate, and to suppress pain

falling sickness, and better than wine, because it cannot rise to the head, as the wine does. (27) Lastly, it is a remedy against a surfeit, for those who are skilful in medicine, when they perceive any man's stomach to be overcome, they first ease it by vomiting, and then, to settle his brain, and to stay the objectionable fumes from ascending to his head, they give him honey on bread. In respect of which great virtues (28) the right composition of those great antidotes, treacle and mithridate[179] - though they consist, the one of more than fifty, the other of more than sixty ingredients – requires three times as much honey as of all the rest. All of which premises considered, no marvel though the wise King said, 'My son, eat honey, for it is good' [180] that the holy land is so often and so much commended for flowing therewith[181] and that the eternal Immanuel used it for his food[182]. Yes, honey, if it is pure and fine is so good in itself, that it must necessarily be good, even for those whose queasy stomachs are against it. But indeed the ordinary honey may well be disliked, for being slovenly handled and greatly spoiled with stopping, and bees both young and old, and some with other mixtures also…

Honey is most suitable for (1) old men, for women and children, for those who are rheumatic and phlegmatic, and generally for all that are of a cold temperature; (2) To young men, and those that are of a hot constitution it is not so good, because it is easily turned into anger, and yet Lobel[183] says we know that honey taken fasting does much good to some natures which have hot livers, and in this point he prefers our English honey…So that he seems to say, that our honey is hurtful to none, because it purges that evil humour, which other honey, in some bodies, is thought to breed. But the proverb says, 'Too much of one thing is good for nothing' and the wise man in his Proverbs, 'It is not good to eat much honey' [184] and also 'Hast thou found honey? Eat as much as is sufficient for you'. For all honey often and immoderately taken (3) causes obstruction; (4) is contrary to its natural quality, and so in time (5) breeds skin disease.

Raw honey (1) loosens the belly, (2) causes coughing, and (3) fills the entrails with wind, especially if it is of the coarser sort. Once boiled it is (4) more nourishing, (5) lighter of digestion, and (6) less laxative in effect, (7) less sharp and purging, for which cause they use it (8) to knit together hollow and crooked ulcers and likewise (9) to close other disjoined flesh. It is also good against (10) pleurisy, against (11) consumption, and all other diseases of the lungs.

Honey is clarified by boiling, either by itself, or else with a fourth part of water, or other liquid. But always in boiling, skim it, that it may be pure.

---

179 A medical preparation of many ingredients
180 Proverbs 24.13
181 Exodus 3 followed by references to Leviticus, Numbers, Deuteronomy.
182 References to Isaiah and Luke follow
183 See note 167
184 Proverbs 25.27

By itself you must boil it until it will yield no more scum (which will take about half an hour) and that with a very low fire, or in a double vessel, lest, by over-heating, it gets a bitter taste, and lest it suddenly runs over and catches alight.

With water it is to be boiled an hour at the least, even until the water is evaporated, which is known by the bubbles that rise from the bottom, then, to make it more pure, put into every pound of honey the white of one egg, and afterwards skim it again in the boiling. The fire may be more fervent at the first, but towards the end it must be slack, for it is then apt to be set on fire, as pure honey, and to become bitter with violent heat.

The coarse honey, once boiled and clarified, has a pleasant taste, and is comparable for most uses to the purest raw bottom-honey.

Which pure honey, if you are disposed to boil it, will need less time to be clarified, yielding little or no scum at all, and in taste and virtue it is more excellent.

When your honey is boiled enough, take it from the fire, sooner rather than later, for if there is any dross remaining, you shall find it in the top, when it is cold, but too much boiling consumes the essence of the honey, and turns the sweet taste into bitter.

And such is honey in its own kind, both raw and boiled. It is also altered by distillation into water, which Raimuadus Lullius[185] that excellent chemist called the quintessence of honey. This quintessence dissolves gold, and makes it potable and likewise any sort of precious stone that is put into it. It is of such virtue, that, if anyone is dying, and drinks two or three drams of it, soon he will revive. If you wash any wound with it, or other sore, it will heal quickly. It is also good against the cough, catarrh and pains of the spleen[186] and against many other diseases. Being given for the space of six and forty days together, to one that has the palsy, it helps him. Which thing John Hester,[187] a practical chemist, in his *Key of Philosophy*, professes himself to have proved. It helps also the falling sickness, and preserves the body from putrefaction. This water is of surprising efficacy.

The making of it is in this manner. Take two pounds of perfect pure honey and put it into a great glass, that four parts of five may remain empty. Lute[188] it well with a head and receiver, and give it fire until there appear certain white fumes, which, by laying wet cloths on the receiver and head, and changing them when they are warm, will turn into a water of a red colour like blood. When it is all distilled, keep the receiver closed shut, and let it stand until it is clear, and of the colour of a ruby. Then distil it in Balneo Maria[189] seven times, and so it will

---

185 Raimundus Lullius c 1233 – c 1315 Spanish Franciscan philosopher & theologian
186 Melt
187 John Hester, English apothecary who translated 'The Key of Philosophy' (1580)
188 To seal a lid with lute, a clay cement, as a protection against fire. Butler adds a note: '*The lute may be made of clay, wool-flocks and salt water, tempered together.*'
189 See note 166

lose this reddish colour, and become yellow as gold, having a great smell and is exceedingly pleasant.

Now as honey is good by itself, either altered or in its own kind, so is it also when mixed with many other plain ingredients, which here to declare would seem but tedious and impertinent. In spite of this, it shall not be amiss, in two or three instances, to give you a taste of such confections, and first of those that are inwardly, then of those that are outwardly received.

Of the first sort are marmalade, and marchpane[190], preserved fruits, such as plum and cherries. Conserves of roses, violets etc with syrups of similar matter.

Marmalade is made in this way. First boil your quinces in their skins until they are soft, then, having pared and strained them, mix with it a similar quantity of clarified honey, and boil this together until it is so thick, that in stirring (for you must continually stir it for fear of burning) you may see the bottom, or, being cooled on a trencher[191], it is thick enough to slice; then take it up and box it speedily. You may also add a quantity of almonds, and nut kernels, also cinnamon, ginger, cloves and mace, of each a similar quantity pounded small and put into the honey with the quinces, and in boiling to be stirred together. This is very good to comfort and strengthen the stomach. For want of quinces you may take wardens, pears, or apples, and especially the pearmain, gillyflower, pippin and royal.

Marchpane may be made in this manner. Boil and clarify by itself, as much honey as you think necessary. When it is cold, take to every pound of honey the white of an egg, and beat them together in a basin, until they are incorporated together and wax white. When you have boiled it again, warm it two or three times on a fire of coals, continually stirring it, then add to it such quantity of blanched almonds or nut kernels, stamped, as shall make it of an appropriate consistency, and after a warm or two more, when it is well mixed, pour it out upon a table, and make up your marchpane. Afterwards you may ice it with rose water and sugar. This is good for the consumption.

Preserve fruits after this manner

The damsons, or other fruit, being gathered fresh from the tree, fair, and in their prime, neither green or sour, nor over-ripe or sweet, with their stalks, but cut short: weigh them, and take their weight in raw fine honey, and adding to the honey a similar quantity of fair water, boil it some half quarter of an hour, or until it yields no scum. Then, having slit the damsons in the dented side for fear of breaking, boil them in this liquid with a low fire, continually skimming and turning them until the meat comes clean from the stone, and then take them up.

---

190 A thick paste made of ground almonds, egg whites, and sugar cooked together, used as an ingredient or eaten on its own as a sweet

191 A flat piece of wood on which meat was served and cut up

If the liquid is then too thin, boil it more; if in the boiling it becomes too thick, put in more fair water, or rose water if you like it. The liquid then being of a suitable consistency, lay up and preserve your fruits in it.

If they are larger fruits, such as quinces, pippins or the like, then is it expedient, when you have bored them through the middle, or have otherwise cored them, to put them in as soon as the liquid is first skimmed, and then to let them boil until they are as tender as cooking apples.

Conserves of roses are to be made in this way: take of the juice of fresh red roses, one ounce; of fine clarified honey, ten ounces and boil these together. When it begins to boil, add four ounces of the leaves of fresh red roses, clipped with scissors into little pieces, boil them until the juice is consumed, and soon put the conserves into some earthen vessel. Keep it in there for a long time, as it will get better and better.

A conserve of violets is made in the same way. Syrup of roses is made like this: steep fresh roses in hot water over the embers (the vessel being covered) until the roses grow pale, then strain out the roses, and put fresh ones in their places, until they also are pale. This do ten times, or until the water is red. And this being purged with egg whites (to every pint of liquid, add one) boil it gently with a similar quantity of fine honey, until it is of convenient thickness. If you prepare it for immediate use, less boiling will serve; if you mean to keep it, it requires more, for which purpose, the sunning of it is good. This purges a little, especially being new.

Or thus. Steep one pound of red rose leaves in four pounds of water, for four and twenty hours. When the water is strained, put into it two pounds of fine honey, and boil it to the thickness of a syrup, taking off the scum as it rises. It tempers the hot affections of the brain, it quenches thirst, it strengthens the stomach, it procures sleep, and stops running eyes and noses.

The syrup of violets is made in the same manner, of fragrant violets, and steeped until the liquid is blue. Being well boiled it may be kept a year without becoming mouldy. It tempers and purges hot and sharp humours, and therefore is good for pleurisy; it expels melancholy, and the effects of that, such as headaches, waking, dreaming, and heaviness of heart; it is suitable to be used before and after purging.

If any man likes better to make these confections with sugar, let him take a similar quantity of honey, for sugar also has with its sweetness a power to preserve, being a kind of honey.

But in respect of the marvellous efficacy, which fine and pure honey has in preserving health, that bulky and earthy stuff is in no way comparable to this celestial nectar. Although some quaint and lady-like palates (whom nothing but that which is far sought and dearly bought can please) unhappily neglect it. In preserving fruits it has more power through its viscosity. Also conserves and syrups being made with honey continue longer, and do more kindly work their effects. So that we may conclude with Ecclesiasticus, 'the bee is little among such as fly, but her fruit is the chief of sweet things.'[192]

Honey is used in external medicines for different purposes, not only to contain the other ingredients in the form of a plaster, but also to open, to clean, to dry, to digest and to resist putrefaction. And therefore it has the predominance in that excellent ointment called *Unguentum Aegyptiacum,* which serves to clean and thoroughly wash old sores, and to take away both dead and protruding flesh. The recipe of which is this. Of verdigris, five ounces, of strong vinegar seven ounces, and of honey fourteen. Boil first the honey and vinegar, and stir them together; after a little while put in the verdigris, being pounded to powder, and then, stirring all together, let them boil until the ointment has appropriate thickness and its purple colour.

Another of similar virtue, but not so corrosive

Boil a quart of good ale in a cooking pot[193], to half a pint, skimming off the froth as it arises. Then put in a spoonful of good honey, and still skimming as is needed, let it boil to the half, or until it is so clammy that being taken up on a stick's end it will not drop, but string down like clarified honey.

What the virtues and properties of meth or hydromel are, may partly be known by that which has been said of honey. For seeing honey is the chief matter of which it is made, it therefore, together with the substance of honey, participates in those natural qualities. Which, by purifying in boiling, together with the benefit of various wholesome ingredients, is rather confirmed and increased, than in any way extenuated or diminished...Meth when it is old is a wine most agreeable to the stomach; it recovers (1) the appetite when lost, it (2) opens the passage of the spirit or breath, it (3) softens the belly, it (4) is good for those who have the cough; (5) if a man takes it, not as his ordinary drink, but, as in medicine, now and then, he shall receive much benefit by it against feverish illnesses, against wasting, and against the diseases of the brain, such as the epilepsy[194], or the falling evil, for which wine is pernicious; it (6) cures the yellow jaundice, it (7) is also good against henbane with milk, and against the winter-cherry, it (8) nourishes the body; (9) so that many have attained to long old age, only by the use thereof. And therefore it was no marvel that Pollio Romulus, who was an hundred years old, imputed the greatest cause of his long continued health to this sovereign drink. (10) For being asked of Augustus the Emperor, by what means especially he had

---

192 Ecclesiasticus 11.3
193 Skillet
194 In describing epilepsy as a 'disease of the brain' Butler shows his debt to medical knowledge.

so long preserved that vigour both of mind and body, his answer was, *'honey inside and outside'.*[195]

The same thing is made clear by the general example of the ancient Britons, who above all other nations, have always been addicted to meth and metheglin. For under heaven there is no fairer people of complexion, nor of more sound and healthy bodies…

And as good and old metheglin excels all wines, as well for pleasantness in taste, as for health, so being distilled it is better than any distilled wine, for comforting and settling of a weak and sick stomach, and for recreating the natural heat.

The manner of distilling it (if you know not) may be this. First set on the fire a deep pot or kettle, almost full of water. When it boils, put in a pewter pot full of metheglin. Before that begins to boil, skim it and put in two or three bruised cloves, and a branch of rosemary, then beat the yolk of an egg in a dish, put into it a spoonful of the meth cold, and stir them together to keep the yolk from curdling. Then put to that a spoonful of the hot meth, and after that another, and another, always beating them together, and then, gradually, put all into the pot, still stirring it about. Then as soon as it boils, take up the pot and, taking care of your hands, pour it into another warm pot of similar capacity, heating it as it runs, and so brew it until it will distil no more. A metheglin posset[196] is of the same virtue.

Wax has no certain, typical quality, but is an average between (1) hot and cold, and between dry and moist. It (2) makes supple the sinews, it (3) ripens and resolves ulcers; (4) the quantity of a piece in wax, being swallowed down by nurses, dissolves the milk curdled in the paps, and (5) ten round pieces of wax, of the size of so many grains of millet or hemp seed, will not allow the milk to curdle in the stomach.

Moreover, it produces the most excellent light, suitable for the uses of the most excellent: for clearness, sweetness, neatness, to be preferred before all others, which Scaliger[197] in his Aenigmata does intimate…This natural yellow wax is by skill, for certain purposes, made white, red and green.

Wax is whitened after this manner. Take the whitest and purest wax, which, being cut into small pieces, put into an earthen vessel, and pour sea-water or brine into it, as much as may suffice to boil it. And cast in also a little saltpetre; all this set over a gentle fire. When it has boiled up twice or thrice, lift the vessel from the fire, and, the wax being soon cooled with cold water, take it out, and when you have scraped off the dross, if any such hang on, and put it into other

---

195 Butler takes this story from Pliny, *Histories*.
196 A drink made from hot milk and curdled with ale, wine, or other liquor, and flavoured with sugar, herbs, spices
197 Scaliger, Julius Caesar (1484-1558), Renaissance author and poet

salt water, and boil it again. And having boiled up twice or thrice, as before, lift it from the fire again. And then take the bottom of another earthen pot, or a little round board with a handle in the middle like a butter churn, but without holes, and having first wetted the bottom of it in cold water, dip it into the hot vessel, and as soon as this wet bottom touches the wax, pull it out again, and you shall have sticking to the bottom a thin cake, which when you have taken it off, wet the bottom again, and dip it as before, and this do until you have taken up all the wax in cakes. These cakes hang in the open air on a line drawn through them, so that they may not touch one another, sprinkling them with water in the sunshine until they are white. If any man would have wax whiter, let him boil it more often, and do all other things in the same manner as before...

To make wax red, take to one pound of wax, in summer, three ounces of clear turpentine, in winter four. These dissolve over a small fire, and by and by take it off to cool a little. Afterwards mix with the red root of anchusa or vermilion, well ground on marble or glass, and sweet oil, of each one ounce. Stir all these and mix them well together. For want of vermilion, use three times as much red lead, but that is not so good.

To make green wax, take instead of vermilion, the same quantity of verdigris.

And such is wax in its kind, both natural and artificial. Natural wax is altered by distillation into an oil of marvellous virtue. Raymond Lully[198] greatly commends it, proving it to be more a celestial or divine medicine than human, because in wounds it works miraculously, which therefore is not much commended by the common surgeon. For it heals a wound, even one which is large and wide, before it is stitched, in the space of eleven days or twelve at the most. But those that are small, this oil heals in three or four days, by anointing only the wound, and laying on a wet cloth in the same. It stops the shedding of the hair, either on the head or beard, by anointing the wounded place.

Also for inward diseases, this oil works miracles. If you give one drachm[199] at a time to drink with white wine, it is excellent in provoking urine which has stopped; it helps stitches and pains in the loins; it helps the cold gout, or sciatica, and all other grievances caused by the cold.

The making or drawing of this oil is done in this way. Take from pure new yellow wax as much as will half fill your retort or glass vessel. Melt it on the fire, and then pour it into sweet wine, in which let it soak; wash it often, and wring it between your hands, then melt it again and pour it into fresh wine; soak it, wash it, and wring it as before, and do this seven times, every time putting it into fresh wine. When you have thus purified the wax, to every pound of it add four ounces

---

198 Lully, Raymond - 13th century Majorcan philosopher and chemist
199 A unit of capacity or volume in the apothecary system, equal to one eighth of a fluid ounce

of the powder of red brick finely bruised, put it all together into your retort of glass, well sealed, then set the retort into an earthen pot, filling it round about and beneath with fine sifted ashes or sand, and set the pot with the body in it on a furnace, and so distil it with a gentle fire. And there will come forth a fair yellow oil, which will congeal in the receiver like mush when it is cold. If you rectify this oil or distil it often, until it will congeal no more, then you will make it too hot to take inwardly, and so lively in the mouth, that you cannot drink it down. As the oil is produced, there will appear in the receiver the four elements: the fire, the air, the water, and the earth, truly astonishing to see.

So excellent is wax by itself, both in its own kind, and when altered by distillation. It is moreover of great use when mixed with others, and is the ground and foundation of waxed cloths and healing ointments, of which to set down two or three examples shall not be amiss.

A cerecloth[200] ...consists chiefly of wax and oil mixed in such proportions, to make the ointment of the right consistency, and therefore (1) when made in summer, or combined with turpentine, lard, gum, marrow, or any liquid thing, a greater quantity of wax is required, and being made in winter, or combined with resin, pitch, metals, dried herbs, powders, or any dry thing, a lesser quantity of wax than oil is convenient.

The ingredients being prepared, first melt the wax, and whatever else of similar nature, such as pitch, suet etc. in the oil over a gentle fire, or in a double vessel, for fear of burning. When they are melted together, put in the powders and other similar ingredients, if there are any, and as soon as you have stirred them well together (before the liquid becomes very hot), set it to cool, and make your cerecloth.

A cerecloth to refresh the wearied sinews and tired muscles is thus to be made. Take (2) oil and wax in equal quantities of two ounces, turpentine two drams, and honey half an ounce.

To comfort the stomach and help concoction, make an ointment in this way. Take (3) oil of mastic[201], of mint, of wormwood, of nutmeg, and small twigs, or any of these, and a convenient quantity of wax.

For the worms in the belly of a child or other, take wax and resin in equal quantities of one ounce, treacle one spoonful, aloes two drams. Melt and mingle the wax and resin together in a pewter dish, on a chafing-dish[202] and coals; being melted, skim it clean, then taking it off, put in the treacle, and stir it together.

---

200 Cloth smeared or impregnated with wax, used as a plaster in surgery
201 An oil made from aromatic gum or resin from the bark of the lentisk or mastic tree
202 A vessel to hold burning charcoal or coal, for heating anything placed upon it

Then having pounded the aloes to powder, strew it, and stir it in, so that it does not clot. And if, by this time, it is too cold to come from the dish, warm it a little upon the chafing-dish again; then having wet the table with butter, pour it on, and work it together with your knife, and so make it up in a roll. To make the dish clean, warm it, and wipe it with a woollen cloth.

This ointment is to be applied to the breast and to the navel. For the navel, spread it on a round piece of leather three inches wide, with a hole in the middle, so that the navel showing through, the plaster may lie both closer and more tightly, and for the breast, spread it on a square piece three inches broad, and twice as long, and lay it upon the breast, between, or close under the nipples.

This do twice together, and let the plasters remain each time on the place, until the heat of the stomach has dried them, and made them loose, which, in those who are much troubled with the worms, will be within twenty four hours, although in some they will stick a whole week together.

For an example of how to make a salve ointment, take a plaster-bandage of *emplastrum de janua*[203], which is marvellously effective in curing green wounds and new ulcers. It assuages inflammation, it cleans, it closes and fills with flesh, and makes whole. It is made in this way: take the juice of parsley, plantain, and betony, in equal amounts of one pound; wax, pitch, resin and turpentine, each of half a pound; boil the wax, pitch, and resin in the juices, softly stirring all together, until the quantity of the juices is reduced, and then taking them off the fire, put in the turpentine, and mix it with the rest.

Another of similar effect

Take deer or mutton suet, wax, resin, in equal amounts of two ounces; turpentine one ounce; boil these together and skim them. Then take this liquid from the fire, and, when it is somewhat cooled, put in two handfuls of the tops of unset hyssop, and stir it about, and setting it over the fire again, boil it softly about a quarter of an hour, until it is green, and then strain it, and let it cool. This is chiefly to be made in May, because then the hyssop is in its prime.

Psalm 111.2:
Great are the works of the Lord,
studied by all who have pleasure in them

---

203 Emplastrum de Janua; or, 'the lesser Plaister of Betony'. Listed in *A Physical directory, or, A translation of the Dispensatory made by the College of Physicians of London* by Nicholas Culpeper (1651)

THE ASTRONOMICAL MONTHS

These are used extensively by Butler to denote seasonal bee-keeping tasks.

Aries	March 21 - April 19
Taurus	April 20 - May 20
Gemini	May 21 - June 20
Cancer	June 21 - July 22
Leo	July 23 - August 22
Virgo	August 23 - September 22
Libra	September 23 - October 22
Scorpio	October 23 - November 21
Sagittarius	November 22 - December 21
Capricorn	December 22 - January 19
Aquarius	January 20 - February 18
Pisces	February 19 - March 20

GLOSSARY

Some of the following words, together with additional ones, are also glossed in the footnotes in the main text.

Blotes Eggs

Bracks Breaches, gaps

Bunt Wheat particle affected by fungus

Bushel One bushel is equivalent to four pecks

Cephens Young drones

Cerecloth Cloth smeared or impregnated with wax or similar substance, and used as a plaster

Chafing-dish A vessel to hold burning charcoal, for heating anything placed on it

Chirurgion Surgeon

Cloom Dried cow dung used to fill cracks and gaps around the skep skirts

Cop A round piece of wood with holes in it to hold together the sticks placed in the hive for internal support

Dams Mothers

Doore/dumbledoore Bumblebee or other large flying insect

Drachm Unit of weight used by apothecaries

Featly Skilfully

Filbert The fruit or nut of the cultivated hazel

Fustian Cloth made of cotton and flax

Gawne/gawn Gallon

Hackle Separate cover made of straw to fit over the skep

Hydromel Drink similar to mead, made with fermented honey and water

Kiver Shallow wooden vessel or tub

Kive Tub or vat for holding liquid

Leere Empty

Litche Bundle

Lozell Rascal, ruffian

Lubbers Idle fellows

Mantle Covering on the skep: '*Your hive being fitted and dressed, you must also have in readiness a mantle, a rest, and a brush. The mantle may be a sheet, or half-sheet, or other linen cloth, an all square at the least.*' Chapter 5.

Mastic, oil of An aromatic gum or resin from the bark of the mastic tree, used in medicine

Metheglin Spiced or medicated mead drink

Moiety Half part

Neat Cow

Nymphs Young female bees

Over-swarmer Box or skep for holding bees left out from a swarm

Peck One peck is equivalent to two gallons

Perch A single perch is a measure of about five and a half yards

Posset Drink made of hot milk curdled with alcohol and often flavoured with spices.

Pottle Pot, tankard, or similar container

Rathe Early, quick

Retort Glass container with a long neck, used in distilling liquids

Sad Sombre colour

Schadons Young bees

Stall, stalle – honey-getting colonies and/or hives

Stool, stools, stooles – the stand on which the hive is placed

Spleetes Small split sticks inserted into hive, to stop the walls collapsing under the weight of honey.

Titmouse Tit

Wicket Door

Withy Willow branch of an osier or other willow, used for tying, binding, or basketry.

Wolmores Gluttons

REFERENCES

Armstrong, Patrick (2000) *The English Parson-Naturalist*. Gracewing.

Bates, A.S. *Charles Butler, Vicar of Wootton, 1600-1647*, pamphlet produced by Charles Butler Memorial Fund, c/o Wootton St Lawrence Parochial Church Council.

Butler, Charles (1623) *The Feminine Monarchie: or the Historie of Bees*. Facsimile, Northern Bee Books, 1985. Text also available online as a PDF document.

Cannadine, David ed. (2004) *The Oxford Dictionary of National Biography*. OUP.

Crane, Eva (1999) *The World History of Beekeeping and Honey Hunting*. Duckworth.

Francis, Leslie J (1989) *The Country Parson*. Gracewing.

Fraser, H Malcolm (1958) *History of Beekeeping in Britain*. Northern Bee Books, reprinted 2010.

Fuller, Thomas (1662) *The Worthies of England*, selected and edited by Richard Barber, Folio Society, London 1987 reprint.

Harding, Joan; Smith, David; Crane, Eva; Duruz, Rosamund M. (1979) *British Bee Books: A Bibliography*. International Bee Research Association.

Hart, A Tindal (1959) *The Country Priest in English History*. Phoenix House.

Hawker, Robert J (2015) *The Earliest Record of Beekeeping in Northern England*. Northern Bee Books.

Hill, Christopher (1998) *England's Turning Point. Essays on 17th century English history*. Bookmarks Publications Ltd.

Hinton, Michael (1994). *The Anglican Parochial Clergy: A Celebration*. SCM Press Ltd.

Pruett, James (1963) *Charles Butler – Musician, Grammarian, Apiarist*. The Musical Quarterly, Volume 49.

Ward, W.R. ed. (1995) *Parson and Parish in Eighteenth-Century Hampshire: Replies to Bishops' Visitations*. Hampshire County Council.

INDEX